Principles and Standards for School Mathematics Navigations Series

NAVIGATING *through* DATA ANALYSIS *and* PROBABILITY *in* GRADES 3–5

Suzanne Chapin
Alice Koziol
Jennifer MacPherson
Carol Rezba

Gilbert J. Cuevas
Grades 3–5 Editor

Peggy A. House
Navigations Series Editor

The National Council of Teachers of Mathematics, Inc.
1906 Association Drive, Reston, VA 20191-1502
(703) 620-9840; (800) 235-7566; www.nctm.org

ISBN 0-87353-521-9

Printed in the United States of America

Table of Contents

Contents of CD-ROM

About This Book

The Data Analysis and Probability Standard for grades 3–5 in *Principles and Standards for School Mathematics* (National Council of Teachers of Mathematics [NCTM] 2000) emphasizes investigations involving data. These investigations should give students opportunities to depict the shape of data sets and use statistical characteristics of the data, such as the range and measures of center, to describe similarities and differences among related sets. In addition, the formulation of conclusions and arguments based on the data should be included in the investigations. The use of appropriate language and symbols should accompany the development of data analysis and simple probability ideas.

The introduction to the volume gives an overview of the development of ideas in data analysis and probability from prekindergarten through grade 12. Each of the four chapters that follow the introduction focuses on a different element of the Data Analysis and Probability Standard. The activities and investigations require students to collect, examine, analyze, and make conclusions about sets of data. Probability activities emphasize the collection of data to develop the idea of probability as a measure of the likelihood of events that are meaningful and real to the students.

In chapter 1, "From Questions to Method: Beginning the Process," the topics include—

- formulating questions that can be addressed with data;
- collecting data using observations, surveys, and experiments; and
- representing data using tables and graphs such as line plots and bar graphs.

Topics in Chapter 2, "Using Data Analysis Methods," include—

- analyzing, summarizing, and describing data sets according to their distributions;
- describing data sets using measures of center; and
- comparing different representations of the same data set.

In chapter 3, "Inferences and Predictions," the topics include—

- making and justifying conclusions and predictions that are based on data.

Topics in Chapter 4, "What Are the Chances?" include—

- exploring the degree of the likelihood of occurrence of certain events; and
- predicting the probability of outcomes of simple experiments and testing the predictions.

Each chapter begins with a discussion of the basic ideas addressed in the topic, followed by student activities and investigations that introduce, and promote familiarity with, the ideas. For each activity, the recommended grade level is identified, and the goals to be achieved, the prerequisite skills and knowledge, and the materials necessary for conducting the activity are presented. Blackline masters, which are

Key to Icons

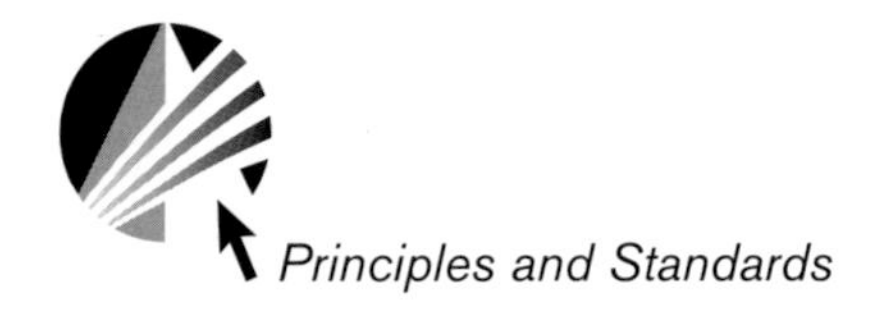

Three different icons appear in the book, as shown in the key. One alerts readers to material quoted from *Principles and Standards for School Mathematics,* another points them to supplementary materials on the CD-ROM that accompanies the book, and a third signals the blackline masters and indicates their locations in the appendix.

signaled by an icon and identified in the materials list, are included and can be found in the appendix. They can also be printed from the CD-ROM that accompanies the book. The CD, also signaled by an icon, contains applets for students to manipulate and resources for professional development.

All the activities have the same format. Each consists of three sections: "Engage," "Explore," and "Extend." The "Engage" section presents tasks designed to address students' interests. "Explore" presents the core investigation that all students should be able to do. "Extend" suggests additional activities intended to expand the ideas that students gained in the core section. Margin notes include teaching tips, anticipated student responses to some of the questions or investigations, references to some of the resources included on the CD-ROM, and quotations from *Principles and Standards for School Mathematics* (NCTM 2000). The assessment section suggests strategies for evaluating students, offers insights about students' performance, and suggests ways to modify the activities for students who are experiencing difficulty or who need enrichment.

As with the other Navigations books, the activities presented are not intended to be a complete curriculum for data analysis and probability in this grade band. Rather, this book presents a collection of activities and investigations that should be used in conjunction with other instructional materials.

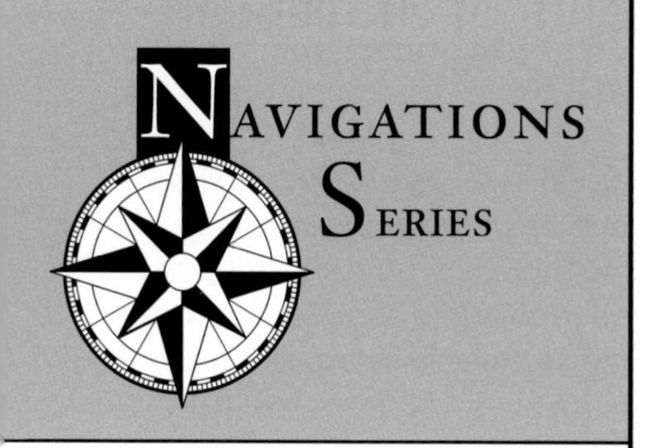

GRADES 3–5

DATA ANALYSIS and PROBABILITY

Introduction

The Data Analysis and Probability Standard in *Principles and Standards for School Mathematics* (NCTM 2000) is an affirmation of a fundamental goal of the mathematics curriculum: to develop critical thinking and sound judgment based on data. These skills are essential not only for a select few but for every informed citizen and consumer. Staggering amounts of information confront us in almost every aspect of contemporary life, and being able to ask good questions, use data wisely, evaluate claims that are based on data, and formulate defensible conclusions in the face of uncertainty have become basic skills in our information age.

In working with data, students encounter and apply ideas that connect directly with those in the other strands of the mathematics curriculum as well as with the mathematical ideas that they regularly meet in other school subjects and in daily life. They can see the relationship between the ideas involved in gathering and interpreting data and those addressed in the other Content Standards—Number and Operations, Algebra, Measurement, and Geometry—as well as in the Process Standards—Reasoning and Proof, Representation, Communication, Connections, and Problem Solving. In the Navigations series, the *Navigating through Data Analysis and Probability* books elaborate the vision of the Data Analysis and Probability Standard outlined in *Principles and Standards.* These books show teachers how to introduce important statistical and probabilistic concepts, how the concepts grow, what to expect students to be able to do and understand during and at the end of each grade band, and how to assess what they know. The books also introduce representative instructional activities that help translate the vision of *Principles and Standards* into classroom practice and student learning.

Fundamental Components of Statistical and Probabilistic Thinking

Principles and Standards sets the Data Analysis and Probability Standard in a developmental context. It envisions teachers as engaging students from a very young age in working directly with data, and it sees this work as continuing, deepening and growing in sophistication and complexity as the students move through school. The expectation is that all students, in an age-appropriate manner, will learn to—

- formulate questions that can be addressed with data and collect, organize, and display relevant data to answer them;
- select and use appropriate statistical methods to analyze data;
- develop and evaluate inferences and predictions that are based on data; and
- understand and apply basic concepts of probability.

Formulating questions that can be addressed with data and collecting, organizing, and displaying relevant data to answer them

No one who has spent any time at all with young children will doubt that they are full of questions. Teachers of young children have many opportunities to nurture their students' innate curiosity while demonstrating to them that they themselves can gather information to answer some of their questions.

At first, children are primarily interested in themselves and their immediate surroundings, and their questions center on such matters as "How many children in our class ride the school bus?" or "What are our favorite flavors of ice cream?" Initially, they may use physical objects to display the answers to their questions, such as a shoe taken from each student and placed appropriately on a graph labeled "The Kinds of Shoes Worn in Kindergarten." Later, they learn other methods of representation using pictures, index cards, sticky notes, or tallies. As children move through the primary grades, their interests expand outward to their surroundings, and their questions become more complex and sophisticated. As that happens, the amount of collectible data grows, and the task of keeping track of the data becomes more challenging. Students then begin to learn the importance of framing good questions and planning carefully how to gather and display their data, and they discover that organizing and ordering data will help uncover many of the answers that they seek. However, learning to refine their questions, planning effective ways to collect data, and deciding on the best ways to organize and display data are skills that children develop only through repeated experiences, frequent discussions, and skillful guidance from their teachers. By good fortune, the primary grades afford many opportunities—often in conjunction with lessons on counting, measurement, numbers, patterns, or other school subjects—for children to pose interesting questions and develop ways of collecting data that will help them formulate answers.

As students move into the upper elementary grades, they will continue to ask questions about themselves and their environment, but their questions will begin to extend to their school or the community or the world beyond. Sometimes, they will collect their own data; at other times, they will use existing data sets from a variety of sources. In either case, they should learn to exercise care in framing their questions and determining what data to collect and when and how to collect them. They should also learn to recognize differences among data-gathering techniques, including observation, measurement, experimentation, and surveying, and they should investigate how the form of the questions that they seek to answer helps determine what data-gathering approaches are appropriate. During these grades, students learn additional ways of representing data. Tables, line plots, bar graphs, and line graphs come into play, and students develop skill in reading, interpreting, and making various representations of data. By examining, comparing, and discussing many examples of data sets and their representations, students will gain important understanding of such matters as the difference between categorical and numerical data, the need to select appropriate scales for the axes of graphs, and the advantages of different data displays for highlighting different aspects of the same data.

During middle school, students move beyond asking and answering the questions about a single population that are common in the earlier years. Instead, they begin posing questions about relationships among several populations or samples or between two variables within a single population. In grades 6–8, students can ask questions that are more complex, such as "Which brand of laundry detergent is the best buy?" or "What effect does light [or water or a particular nutrient] have on the growth of a tomato plant?" They can design experiments that will allow them to collect data to answer their questions, learning in the process the importance of identifying relevant data, controlling variables, and choosing a sample when it is impossible to collect data on every case. In these middle school years, students learn additional ways of representing data, such as with histograms, box plots, or relative-frequency bar graphs, and they investigate how such displays can help them compare sets of data from two or more populations or samples.

By the time students reach high school, they should have had sufficient experience with gathering data to enable them to focus more precisely on such questions of design as whether survey questions are unambiguous, what strategies are optimal for drawing samples, and how randomization can reduce bias in studies. In grades 9–12, students should be expected to design and evaluate surveys, observational studies, and experiments of their own as well as to critique studies reported by others, determining if they are well designed and if the inferences drawn from them are defensible.

Selecting and using appropriate statistical methods to analyze data

Teachers of even very young children should help their students reflect on the displays that they make of the data that they have gathered. Students should always thoughtfully examine their representations to determine what information they convey. Teachers can prompt young children to derive information from data displays through questions like "Do more children in our class prefer vanilla ice cream, or do more prefer chocolate ice cream?" As children try to interpret their work, they come to realize that data must be ordered and organized to convey answers to their questions. They see how information derived from data, such as their ice cream preferences, can be useful—in deciding, for example, how much of particular flavors to buy for a class party. In the primary grades, children ordinarily gather data about whole groups—frequently their own class—but they are mainly interested in individual data entries, such as the marks that represent their own ice cream choices. Nevertheless, as children move through the years from prekindergarten to grade 2, they can be expected to begin questioning the appropriateness of statements that are based on data. For example, they may express doubts about such a statement as "Most second graders take ballet lessons" if they learn that only girls were asked if they go to dancing school. They should also begin to recognize that conclusions drawn about one population may not apply to another. They may discover, for instance, that bubble gum and licorice are popular ice cream flavors among their fellow first graders but suspect that this might not necessarily be the case among their parents.

In contrast with younger children, who focus on individual, often personal, aspects of data sets, students in grades 3–5 can and should be guided to see data sets as wholes, to describe whole sets, and to compare one set with another. Students learn to do this by examining different sets' characteristics—checking, for example, values for which data are concentrated or clustered, values for which there are no data, or values for which data are unusually large or small (*outliers*). Students in these grades should also describe the "shape" of a whole data set, observing how the data spread out to give the set its *range*, and finding that range's center. In grades 3–5, the center of interest is in fact very often a measure of a data set's center—the *median* or, in some cases, the *mode*. In the process of learning to focus on sets of data rather than on individual entries, students should start to develop an understanding of how to select *typical* or *average* (*mean*) values to represent the sets. In examining similarities and differences between two sets, they should explore what the means and the ranges tell about the data. By using standard terms in their discussions, students in grades 3–5 should be building a precise vocabulary for describing the characteristics of the data that they are studying.

By grade 5, students may begin to explore the concept of the mean as a balance point in an informal way, but a formal understanding of the mean and its use in describing data sets does not become important until grades 6–8. By this time, just being able to compute the mean is no longer enough. Students need ample opportunities to develop a fundamental conceptual understanding—for example, by comparing the usefulness and appropriateness of the mean, the median, and the mode as ways of describing data sets in different contexts. In middle school, students should also explore questions that are more probing, such as "What impact does the spread of a distribution have on the value of the mean [or the median]?" Or "What effect does changing one data value [or more than one] have on different measures of center—the mean, the median, and the mode?" Technology, including spreadsheet software, calculators, and graphing software, becomes an important tool in grades 6–8, enabling students to manipulate and control data while they investigate how changes in certain values affect the mean, the median, or the distribution of a set of data. Students in grades 6–8 should also study important characteristics of data sets, such as *symmetry*, *skewness*, and *interquartile range*, and should investigate different types of data displays to discover how a particular representation makes such characteristics more or less apparent.

As these students move on into grades 9–12, they should grow in their ability to construct an appropriate representation for a set of univariate data, describe its shape, and calculate summary statistics. In addition, high school students should study linear transformations of univariate data, investigating, for example, what happens if a constant is added to each data value or if each value is multiplied by a common factor. They should also learn to display and interpret bivariate data and recognize what representations are appropriate under particular conditions. In situations where one variable is categorical—for example, gender—and the other is numerical—a measurement of height, for instance—students might use appropriately paired box plots or histograms to compare the heights of males and females in a given group. By contrast, students who are presented with bivariate numerical data—for example, measurements of height and arm span—might use a scatterplot to represent their data, and they should be able to describe the shape of the scatterplot and use it to analyze the relationship between the two lengths measured—height and arm span. Types of analyses expected of high school students include finding functions that approximate or "fit" a scatterplot, discussing different ways to define "best fit," and comparing several functions to determine which is the best fit for a particular data set. Students should also develop an understanding of new concepts, including *regression*, *regression line*, *correlation*, and *correlation coefficient*. They should be able to explain what each means and should understand clearly that a correlation is not the same as a causal relationship. In grades 9–12, technology that allows users to plot, move, and compare possible regression lines can help students develop a conceptual understanding of residuals and regression lines and can enable them to compute the equation of their selected line of best fit.

Developing and evaluating inferences and predictions that are based on data

Observing, measuring, or surveying every individual in a population is an appropriate way of gathering data to answer selected questions. Such "census data" is all that we expect from very young children, and teachers in the primary grades should be content when their students confine their data gathering and interpretation to their own class or another small group. But as children mature, they begin to understand that a principal reason for gathering and analyzing data is to make inferences and predictions that apply beyond immediately available data sets. To do that requires sampling and other more advanced statistical techniques.

Teachers of young children lay a foundation for later work with inference and prediction when they ask their students whether they think that another group of students would get the same answers from data that they did. After discussing the results of a survey to determine their favorite books, for example, children in one first-grade class might conclude that their peers in the school's other first-grade class would get similar results but that the fourth graders' results might be quite different. The first graders could speculate about why this might be so.

As students move into grades 3–5, they should be expected to expand their ability to draw conclusions, make predictions, and develop arguments based on data. As they gain experience, they should begin to understand how the data that they collect in their own class or school might or might not be representative of a larger population of students. They can begin to compare data from different samples, such as several fifth-grade classes in their own school or other schools in their town or state. They can also begin to explore whether or not samples are representative of the population and identify factors that might affect representativeness. For example, they could consider a question like "Would a survey of children's favorite winter sports get similar results for samples drawn from Colorado, Hawaii, Texas, and Ontario?" Students in the upper grades should also discuss differences in what data from different samples show and factors that might account for the observed results, and they can start developing hypotheses and designing investigations to test their predictions.

It is in the middle grades, however, that students learn to address matters of greater complexity, such as the relationship between two variables in a given population or sample, or the relationships among several populations or samples. Two concepts that are emphasized in grades 6–8 are *linearity* and *proportionality*, both of which are important in developing students' ability to interpret and draw inferences from data. By using scatterplots to represent paired data from a sample—for example, the height and stride length of middle schoolers—students might observe whether the points of the scatterplot approximate a line, and if so, they can attempt to draw the line to fit the data. Using the slope of that line, students can make conjectures about a relationship between height and stride length. Furthermore, they might decide to compare a scatterplot for middle school boys with one for middle school girls to determine if a similar ratio applies for both groups. Or they might draw box plots or relative-frequency histograms to represent data on the heights of samples of middle school boys and high school boys to investigate the variability in height of boys of different ages. With the help of graphing technology, students can examine many data sets and learn to differentiate between linear and nonlinear relationships, as well as to recognize data sets that exhibit no relationship at all. Whenever possible, they should attempt to describe observed relationships mathematically and discuss whether the conjectures that they draw from the sample data might apply to a larger population. From such discussions, students can plan additional investigations to test their conjectures.

As students progress to and through grades 9–12, they can use their growing ability to represent data with regression lines and other mathematical models to make and test predictions. In doing so, they learn that inferences about a population depend on the nature of the samples, and concepts such as *randomness, sampling distribution*, and *margin of error* become important. Students will need firsthand experience with many different statistical examples to develop a deep understanding of the powerful ideas of inference and prediction. Often that experience will come through simulations that enable students to perform hands-on experiments while developing a more intuitive understanding of the relationship between characteristics of a sample and the corresponding characteristics of the population from which the sample was drawn. Equipped with the concepts learned through simulations, students can then apply their understanding by analyzing statistical inferences and critiquing reports of data gathered in various contexts, such as product testing, workplace monitoring, or political forecasting.

Understanding and applying basic concepts of probability

Probability is connected to all mathematics from number to geometry. It has an especially close connection to data collection and analysis. Although students are not developmentally ready to study probability in a formal way until much later in the curriculum, they should begin to lay the foundation for that study in the years from prekindergarten to grade 2. For children in these early years, this means informally considering ideas of likelihood and chance, often by thinking about such questions as "Will it be warm tomorrow?" and realizing that the answer may depend on particular conditions, such as where they live or what month it is. Young children also recognize that some things are sure to happen whereas others are impossible, and they begin to develop notions of *more likely* and *less likely* in various everyday contexts. In addition, most children have experience with common devices of chance used in games, such as spinners and dice. Through hands-on experience, they become aware that certain numbers are harder than others to get with two dice and that the pointer on some spinners lands on certain colors more often than on others.

In grades 3–5, students can begin to think about probability as a measurement of the likelihood of an event, and they can translate their earlier ideas of *certain, likely, unlikely*, or *impossible* into quantitative representations using 1, 0, and common fractions. They should also think about events that are neither certain nor impossible, such as getting a 6 on the next roll of a die. They should begin to understand that although they cannot know for certain what will happen in such a case, they can associate with the outcome a fraction that represents the frequency with which they could expect it to occur in many similar situations. They can also use data that they collect to estimate probability—for example, they can use the results of a survey of students' footwear to predict whether the next student to get off the school bus will be wearing brown shoes.

Students in grades 6–8 should have frequent opportunities to relate their growing understanding of proportionality to simple probabilistic situations from which they can develop notions of chance. As they refine their understanding of the chance, or likelihood, that a certain event will occur, they develop a corresponding sense of the likelihood that it will not occur, and from this awareness emerge notions of complementary events, mutually exclusive events, and the relationship between the probability of an event and the probability of its complement. Students should also investigate simple compound events and use tree diagrams, organized lists, or similar descriptive methods to determine probabilities in such situations. Developing students' understanding of important concepts of probability—not merely their ability to compute probabilities—should be the teacher's aim. Ample experience is important, both with hands-on experiments that generate empirical data and with computer simulations that produce large data samples. Students should then apply their understanding of probability and proportionality to make and test conjectures about various chance events, and they should use simulations to help them explore probabilistic situations.

Concepts of probability become increasingly sophisticated during grades 9–12 as students develop an understanding of such important ideas as *sample space*, *probability distribution*, *conditional probability*, *dependent* and *independent events*, and *expected value*. High school students should use simulations to construct probability distributions for sample spaces and apply their results to predict the likelihood of events. They should also learn to compute expected values and apply their knowledge to determine the fairness of a game. Teachers can reasonably expect students at this level to describe and use a sample space to answer questions about conditional probability. The solid understanding of basic ideas of probability that students should be developing in high school requires that teachers show them how probability relates to other topics in mathematics, such as counting techniques, the binomial theorem, and the relationships between functions and the area under their graphs.

Developing a Data Analysis and Probability Curriculum

Principles and Standards reminds us that a curriculum that fosters the development of statistical and probabilistic thinking must be coherent, focused, and well articulated—not merely a collection of lessons or activities devoted to diverse topics in data analysis and probability. Teachers should introduce rudimentary ideas of data and chance deliberately and purposefully in the early years, deepening and expanding their students' understanding of them through frequent experiences and applications as students progress through the curriculum. Students must be continually challenged to learn and apply increasingly sophisticated statistical and probabilistic thinking and to solve problems in a variety of school, home, and real-life settings.

The six *Navigating through Data Analysis and Probability* books make no attempt to present a complete, detailed data analysis and probability curriculum. However, taken together, these books illustrate how selected "big ideas" behind the Data Analysis and Probability Standard develop this strand of the mathematics curriculum from prekindergarten through grade 12. Many of the concepts about data analysis and probability that the books present are closely tied to topics in algebra, geometry, number, and measurement. As a result, the accompanying activities, which have been especially designed to put the Data Analysis and Probability Standard into practice in the classroom, can also reinforce and enhance students' understanding of mathematics in the other strands of the curriculum, and vice versa.

Because the methods and ideas of data analysis and probability are indispensable components of mathematical literacy in contemporary life, this strand of the curriculum is central to the vision of mathematics education set forth in *Principles and Standards for School Mathematics*. Accordingly, the *Navigating through Data Analysis and Probability* books are offered to educators as guides for setting successful courses for the implementation of this important Standard.

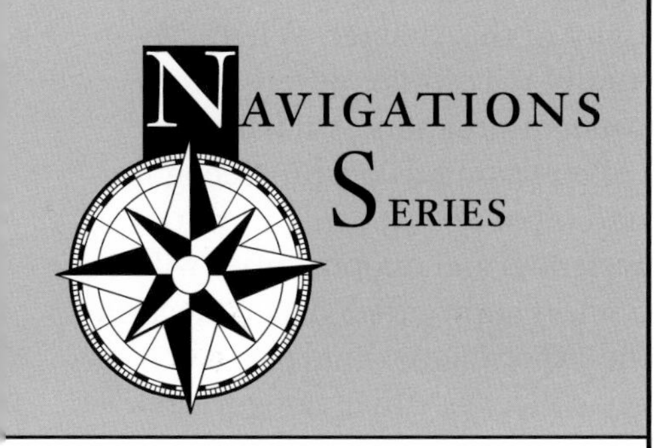

DATA ANALYSIS *and* PROBABILITY

GRADES 3–5

Chapter 1

From Questions to Method: Beginning the Process

Can you predict how many students in your school will eat dinner at a fast-food restaurant this week? Can you estimate how much money will be spent in the United States on CDs in the next three years? What movie was the top revenue producer for Hollywood during the decade of the 1990s? What is the average cost of a movie ticket in the United States? In today's rapidly changing information age, students are constantly being called on to answer questions that are based on an analysis of data. The results of surveys, studies, experiments, and polls are commonplace in newspapers and magazines, on the nightly news, and on the Internet.

Students who develop skills in analyzing data in meaningful ways will be prepared to formulate arguments based on quantitative information. In grades 3–5, we need to help students continue their development of the use of data analysis tools by offering them a variety of experiences in posing questions so they can collect, organize, and display data that address the questions.

The process of data analysis begins with the formulation of questions concerning an issue or topic of interest. Students should be encouraged to formulate questions that address issues in their everyday lives at school, at home, or within their communities. All data investigations begin with questions, yet asking good questions is a skill that takes time to develop. The first activity in this chapter, Questions, Please? engages students in developing an understanding of the components of a well-conceived investigation. Students will find that posing appropriate questions involves an iterative process requiring some preliminary data collection and exploration.

It is important for students to collect data from real-world sources in their investigations. In the second activity of this chapter, What's My Method? students will have an opportunity to consider different formats for data collection. In real-life problem situations, students will need to decide on appropriate means of collecting data while recognizing that data are sometimes "messy" and often do not easily fit into neatly organized formats. Guided experiences and opportunities will help students develop skills needed to select appropriate data collection tools, such as observations, surveys, and experiments. Students will use the vocabulary to describe the data they collect as either categorical data or numerical data. Students will also recognize the appropriate graphical representation to use for the types of data they have collected.

Questions, Please?

Grades 3–5

The first step in data analysis involves posing appropriate questions to frame the scope and direction of the work. In this activity, students design a question for data analysis by using a three-step process. First, students identify a meaningful topic or problem context by using a list of issues, themes, and interests. Second, students pose questions to indicate the purpose of their data analysis, which may—

- describe or summarize a set of data;
- determine preferences and opinions from a set of data;
- compare and contrast two or more sets of data; or
- generalize or make predictions from a set of data.

Finally, using a given set of broad discussion guidelines, students work in pairs to refine the question they have posed for a data investigation.

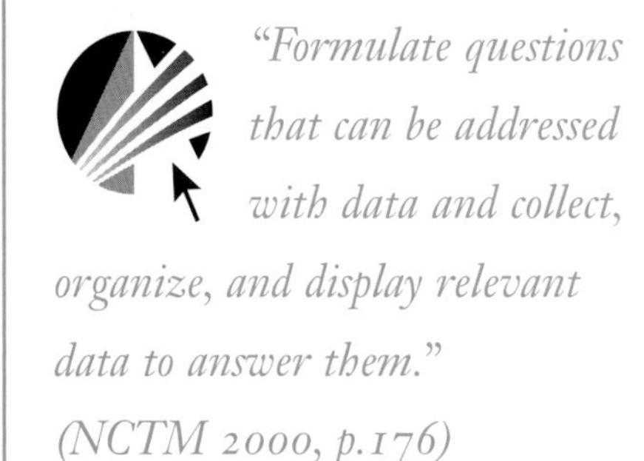

"Formulate questions that can be addressed with data and collect, organize, and display relevant data to answer them." (NCTM 2000, p.176)

Goals

Students will—

- pose questions that reflect a topic and purpose that they have identified for a data investigation; and
- refine their questions to be addressed by data collection and analysis.

The following three-step procedure can be used to formulate a data analysis question:

1. Identify a topic from a list of subjects, themes, and interests.
2. Apply a question stem to indicate the purpose of the data investigation.
3. Refine the question posed for a data analysis.

Prior Knowledge

By grade 3, students should have had experiences in formulating questions about a favorite activity. For example, students will have posed questions about themselves, such as what is their favorite sport, their favorite ice-cream flavor, or their favorite type of pet. By grade 4, students should have collected and represented data in graphs and possibly analyzed data to answer questions.

Materials and Equipment

- A transparency of the blackline masters "Determining a Purpose for a Data Investigation" and "Getting Ready"
- Copies of the blackline masters "Getting Ready" and "A Question to Investigate" for each pair of students
- An overhead projector

pp. 88, 92, 94

Classroom Environment

Students work in pairs to carry out the activity.

Activity

Engage

Tell the students that they will be formulating questions for their own data investigation. Indicate that each question should have a purpose. Place a transparency of the blackline master "Determining a

Purpose for Data Investigation" on the overhead projector. Highlight the purposes as you direct students to review the list of four purposes or reasons for conducting an investigation:

1. To describe or summarize what was learned from a set of data
2. To determine preferences and opinions from a set of data
3. To compare and contrast two or more sets of data
4. To generalize or make predictions from a set of data

As you discuss each purpose, have the students look at the set of related question stems and sample questions that may be used to formulate questions to achieve the identified purpose.

Distribute handouts of the blackline master "Getting Ready" to pairs of students and place a transparency of the same blackline master on the overhead. Tell the students that there is a sample graph for each of the four data analysis purposes that they have discussed. Select one of the graphs, reviewing with the students the title of the graph and the purpose of the data investigation. Ask the students in each pair to work together to identify a question that may have been posed to achieve the purpose of the data investigation. Give students time to discuss the possible questions before asking them to share the questions with the class. Help them achieve consensus on a question that could have been posed for the selected data investigation.

Possible questions posed for Getting Ready:

1. What are the greatest and least numbers of letters found in the fifth graders' first names?
2. What are students' most and least favorite types of cooked potatoes?
3. How do third and fifth graders differ in their favorite sodas?
4. How fast can girls run a mile? Compare the times for girls of different ages.

Have students continue working in pairs, repeating the process of identifying appropriate questions for the remaining three graphs on the blackline master "Getting Ready." Review the results of each pair's work, emphasizing reaching consensus on the questions that might have been posed for each data investigation. Students' answers may vary slightly from those indicated in the answer key.

Explore

Arrange the students in pairs. Give one copy of the blackline master "A Question to Investigate" to each pair of students and explain that they will use it as a guide in formulating a question for a data investigation.

Direct students to the first section of "A Question to Investigate," which lists subjects, themes, and interests. Indicate that these are some broad areas from which they might identify a topic. Read through the list and ask students to elaborate on some of the topics that might be embedded in the broad ones. Some examples are given in figure 1.1.

From the list of subjects, themes, and interests, first have students select one and expand on it to identify the topic they wish to use in formulating their question. They should record their selections on the sheet. Second, have the students determine the purpose of their data investigation and suggest that they use one of the related question stems to help them formulate questions about the topic. Model one example for the students as follows:

Interest:	Books
Topic:	Favorite types of books read by students
Purpose:	To determine preferences and opinions from a set of data

Formulated question: What are the favorite types of books read by students in our class?

Subjects/Themes/Interests	Sample Topics
Environment	Bird population in an area; preservation of the rain forest; effects of global warming on oceans
National disasters	Number and severity of earthquakes, tornadoes, or hurricanes in an area
Pollution	Smog count during different times of the year
Health habits	Number of hours of exercise per week; washing hands before meals; brushing teeth
Littering	Paper found on streets, roads, or classroom floors
Nutrition habits	Number of fruits and vegetables that students eat per day
Second-hand smoke	Number of students living with smokers
Highway safety	Number of people riding bicycles on the correct side of the road and number of people riding bicycles on the wrong side of the road
Childhood diseases	Number of students who have had measles, mumps
Waste	What students throw away at lunch

Fig. **1.1.**
Examples for data investigation

The World Almanac *and the* Guinness Book of World Records *are great sources of ideas for data investigation topics.*

Third, allow time for each pair of students to formulate their question and record it at the bottom of the blackline master. Fourth, have each pair discuss their formulated question with another pair. During this discussion, each pair should be directed to interview the other pair, asking questions that could further refine or limit their question to ensure that an answer to the question will be feasible through data collection. For example, in the interview, each pair of students should first ask the other pair what answer they would give to their question. This should assist them in checking for both the clarity and conciseness of the question. If the response is very different from the expected response options, the questioners should ask the other pair to tell them what they heard when asked the question or how they interpreted the question that was asked. From their responses, the first pair should begin to decide whether the question they formulated will elicit the information they want to collect. At the end of the interviews, each pair should be given time to refine their question. Finally, have each pair state to the class the question that they have formulated for their data investigation. Post the questions on a bulletin board.

Extend

Ask the students to specify the data they will need to collect to answer the question they have formulated. For example, for the sample topic in the "Explore" section (favorite types of books read by students), students might identify the possible categories of books read by students before collecting the data. In this example, the data might include such categories as mysteries, science fiction, biographies, classics, and so on. To ensure that the categories selected were mutually exclusive and did not overlap, students might wish to check with the school librarian or some other knowledgeable source or reference.

Assessment Ideas

When assessing whether students can formulate a question that reflects an identified topic and purpose for a data investigation, pay particular attention to whether the question addresses the purpose within the context of the topic. This task requires students to synthesize two parts into a whole, merging the topic and the purpose into one formulated question. Some students at this age may need help merging the two ideas. You want students to recognize that each data collection has a purpose and that by including the purpose in their formulation of their question, they will be able to base their analysis of the data that they subsequently collect on that purpose.

Where to Go Next in Instruction?

As students begin to gather data to investigate a question that they have formulated, they need to be aware of the variety of methods that can be used to collect data for an investigation. The next activity, What's My Method? helps students think about how they can collect data for the questions they pose.

What's My Method?

Grades 4–5

The activity guides students through the development of a plan to collect the data necessary to answer an investigation question. A variety of data collection methods are introduced: observations, surveys, experiments, measurements, interviews, polls. In addition, students will be directed to obtain data from existing sets such as census or historical data, to perform a simulation, or to use the Internet. This activity also helps students to develop an understanding that there are many ways to collect data.

"Collect data using observations, surveys, and experiments." (NCTM 2000, p. 176)

Goals

Students will—

- discuss data collection methods;
- identify and describe appropriate data collection methods to answer a set of questions; and
- determine whether the data to be collected will be categorical or numerical.

An example of categorical data would be information on students' preferred flavors of ice cream; numerical data could be information on their heights.

Prior Knowledge

Students should have had opportunities and experiences in data collection and analysis. They should know how to organize data into charts and tables and how to represent data in pictographs and simple bar graphs. They should be able to notice individual aspects of the data, such as the value that occurs most frequently or is the greatest or least value. They should also have some understanding of how the data were collected—that is, through a count like a show of hands, by collecting objects, through tallies, or through a simple one-question survey.

A *pictograph* uses pictures to show and compare information.

A *bar graph* uses separate bars of different heights or lengths to show and compare data.

Materials and Equipment

- Transparencies of the blackline masters "What's My Method?—Descriptions" and "What's My Method?—Explorations"
- A copy of the blackline master "Data Sets" for each group of four students
- A copy of the blackline master "What's My Method?—Exploration" for each pair of students
- An index card for each pair of students

pp. 96, 98, 100

Classroom Environment

For the first part of the activity, students work in groups of four. They work in pairs to answer the questions in the second part.

Activity

Engage

Involve the students in a review of prepared data sets to identify possible methods of data collection. Distribute copies of the blackline master "Data Sets" to each group of four students and ask them to select a

Data collection methods include the following:

- Observations
- Surveys and questionnaires
- Experiments
- Interviews
- Polls
- Examinations of past records
- Simulations
- Searches of the Internet, library, or other resources

recorder and a reporter. The role of the recorder will be to write down each of the answers to the question suggested by members of the group, and the role of the reporter will be to report to the class on the group's responses. Indicate to the class that they will have five minutes to generate answers to the question.

The blackline master "Data Sets" presents different displays of data that answer the following questions:

- What are students' favorite types of pancakes?
- How many votes did the candidates for class president receive in the student elections?
- What are the attributes of the dogs in the dog database?
- What was the lowest temperature recorded for each day during the month of October?
- Over the six-year period from 1990–1995, how many hurricanes were reported in total for each of the months from June through November?

Discuss these questions with the class or ask the groups of students to examine the displays and form the questions on their own. Then pose the following situation to the class: "Suppose you want to investigate topics to answer questions similar to those answered by the graphs on the activity sheet 'Data Sets.' List all the different ways in which you could gather the data (or the different tools that you could use) to answer the questions."

Students may need suggestions to get started, and you may wish to give them a clue that is related to something familiar. For example, if the students recently participated in a class election, you could ask the class what method was used to collect data in the election. You might tell the groups the answer (e.g., a poll, or an election poll) or leave them to come up with it as one of their responses to the question.

After five minutes, have each group share one of the items on its list in a round-robin fashion; have the first group's reporter identify one method, ask the next group to report on a method that hasn't already been shared, and continue around the room until all the groups have tried to report an answer. As each group reporter reads off one of his or her group's responses, you should record it on chart paper, a transparency, or the chalkboard. Groups should report only a method that has not already been given by another group. If a group does not have a new entry to add to the list, they should yield to the next group.

When all groups have had an opportunity to report once, continue for a second round, beginning with the first group. Repeat the round-robin process until all groups have shared the data collection methods they identified. All reasonable responses, similar to the entries on the list found in the margin, should be recorded on the master list. You should also add and explain any other methods of data collection that were not presented by the groups.

As each method is identified, ask the students to describe the method and how the method is similar to, and different from, other methods previously mentioned. Clarify the description of these methods by displaying a transparency of "What's My Method?—Descriptions" or by giving a copy of the blackline master to each student.

Explore

Arrange the students in pairs and distribute a copy of the activity page "What's My Method?—Explorations" to each pair. Students should discuss the method that they would use to collect data for these fourteen data investigations listed. You should have students identify and describe the method from the list, which includes observations, surveys and questionnaires, polls, interviews, experiments, measuring, examining past records, simulations, and searching the Internet, library, or other resources.

When the students have completed this task, hold a class discussion about their choices. In this discussion, you may find that students have chosen more than one way to collect the data; their choices will usually reflect the ways in which they have interpreted the question.

Second, discuss the difference between categorical data and numerical data, the two types of data that students collect during data investigations. *Categorical data* represent individuals or objects by one or more characteristics or traits that they share, such as maleness or femaleness or blue eyes or green eyes. Categories are nonoverlapping classifications, and categorical data are often treated as counts, proportions, or percentages of the people or things in them. *Numerical data* represent objects or individuals by numbers assigned to certain measurable properties, such as their length or their age. Use the following examples to help in pointing out the differences between the two types of data.

- Categorical data—favorite ice-cream flavors; hair color.
- Numerical data—ages; number of students in each class in a school

You can check for understanding of these two types of data by playing a little game, "Name That Data Type!" Have students identify whether the data you select are categorical (C) or numerical (N). You can use examples like the following: lengths of students' strides (N); students' eye color (C); students' heights (N); ways in which students like to travel (C); students' arm spans (N); zoo animals (C); students' weights at birth (N); the number of siblings that students have (N); crayon colors (C); bean-plant growth (N); favorite color of apples (C); favorite holidays (C); and so on. Continue the game by having the students create questions to challenge their classmates.

Ask the students what type of data they will collect when conducting their data investigation and why they think it is the type they name. Pay close attention to the reasons students give to support what they think the type might be.

Extend

Have pairs of students take the question that they formulated in the activity Questions, Please? on the blackline master "A Question to Investigate" or have them pose a question for which they can collect data by using one of the methods discussed in this activity. Have each pair design a data collection tool (e.g., a survey or questionnaire, experiment, etc.) and indicate the type of data (i.e., categorical or numerical) that they will collect. When the data collection tool is complete, have each pair exchange their work with another pair of students to get feedback. Pairs should review the data collection tool by critiquing it with respect to the question that has been posed and the type of data that will be collected, determining whether these are reasonable and

Answers to What's My Method?–Explorations may vary. Suggested answers follow:

1. SQ
2. O
3. S
4. SQ
5. E
6. P
7. E
8. O
9. S
10. R
11. S
12. I
13. E
14. SQ

feasible, and making suggestions for adjustments where appropriate. Once the students have made any recommended revisions, you should review the data collection tool, making your own recommendations for revisions. Continue this activity by giving each pair the opportunity to collect their data using their data collection instrument.

Assessment Ideas

To assess students, you might use the following process, in which pairs of students determine the type of data collection method that they would use for a problem they select, and then they try to identify the data collection methods appropriate for the questions formulated by the other pairs. Have each pair of students formulate a question for a data investigation by following the procedures described in the first activity. Each pair of students should record its question on an index card that you supply. Before distributing an index card to each pair, number the back of each card, beginning with 1. Distribute the cards, and ask students to write down the question on the front of the numbered index card. Tell them to describe the method that they would use for collecting data and record this next to their number on a numbered master list that you will keep at your desk. For example, the pair of students with index card number 4 will write its data collection method (e.g., experiment) on the answer master next to number 4.

Once all pairs have written their questions and recorded their methods on the master list, have each pair number a sheet of paper up to the number of pairs in the class. Shuffle the index cards and pass them out to the pairs of students. Once the students have determined the method that would be used to gather the data to answer the numbered question, have the students record the method on their paper. Then, in an organized manner, each pair of students should pass its index card to another pair of students, receiving in turn an index card from a third pair of students. When all pairs have read all the index cards and identified a method they would use for data collection for each situation, collect the papers and assess the pairs' responses. As you evaluate their answers, assess their ability to formulate questions as well as their ability to determine an appropriate method for gathering data for the data investigation.

For additional ideas on this topic, see Bohan, Irby, and Vogel (1995), Feicht (1999), Hitch and Armstrong (1994), and Parker and Widner (1992), on the CD-ROM.

Where to Go Next in Instruction?

Once students become comfortable posing questions about topics of interest and understand the different methods that can be used to collect data, they need to explore data analysis strategies. In chapter 2, students will learn to select and use appropriate methods to analyze the data they collect.

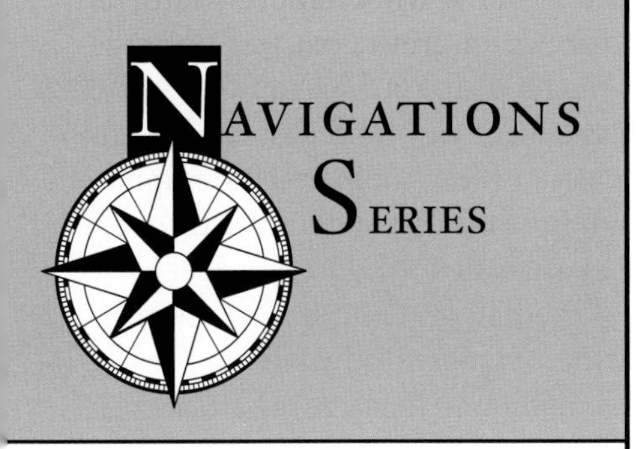

DATA ANALYSIS *and* PROBABILITY

GRADES 3–5

Chapter 2
Using Data Analysis Methods

The investigations in this chapter illustrate ways to develop students' understanding of methods for analyzing sets of data. Each investigation demonstrates approaches to help students learn how to analyze, summarize, and describe data sets, first by learning how to examine the shape of the data and then by learning about the concepts of central tendency. Each investigation emphasizes learning how to describe a set of data according to its distribution. Special attention is given to developing students' understanding of *representativeness*—that there is a value in a data set that can be used to describe or indicate what is typical. Building on students' understanding of "most" and "middle," the investigations emphasize learning to describe what is typical of a data set and introduce students to the concepts of mode, median, and mean.

The first two investigations, Long Jump and How Many Stars Can You Draw in One Minute? involve students in collecting data and then describing the shape and features of data sets. The activities emphasize learning how to summarize data by considering the typical value in the data set. These investigations present opportunities to introduce the terms *range*, *outlier*, *clusters*, *gaps*, and *typical*. The first investigation also introduces students to the use of line plots as tools for organizing, recording, and analyzing data. Students also use line plots in the second investigation and begin to make comparisons of data sets by examining similarities and differences in the shape of the data on line plots.

"Describe the shape and important features of a set of data and compare related data sets, with an emphasis on how the data are distributed." *(NCTM 2000, p. 176)*

Part of learning how to analyze data involves developing a conceptual understanding of the three types of averages, or measures of central tendency: mode, median, and mean. Students need to know more than

"Use measures of center, focusing on the median, and understand what each does and does not indicate about the data set."
(NCTM 2000, p 176)

the procedures for finding each. They need to know what each statistic tells and does not tell and how to interpret each in the context of the entire data set. The last two activities in the chapter, Do You Get Enough Sleep? and Exploring the Mean, focus on the measures of central tendency. As with the first two investigations, each activity emphasizes collecting, organizing, describing, and summarizing data. In both, students continue to use line plots as an analysis tool. Do You Get Enough Sleep? explores the concept of median and emphasizes comparing data sets. The fourth investigation, Exploring the Mean, presents an approach for introducing the concept of mean as a balance point in a distribution.

Long Jump

Grades 3–4

This investigation introduces students to the use of a line plot as a tool for recording, organizing, and analyzing data. As students investigate how far they can jump, they begin to learn how to summarize and describe a distribution by using terms like *range*, *cluster*, *gap*, and *outlier*.

Students collect data for each member of the class and compile these data on a line plot. Then students describe the data distribution and use it to decide the length of the typical jump in the class. If all students are not able to participate in a jumping activity, you can alter this investigation by changing the data that the students will collect. Instead of having them measure the length of a jump and investigate the typical jump length, you might ask them to collect data within another context. For example, students might measure their arm spans or the distance they can blow a paper clip across a table with one breath.

"Describe the shape and important features of a set of data and compare related data sets, with an emphasis on how the data are distributed." *(NCTM 2000, p. 176)*

Goals

Students will—

- make predictions about the jumping ability of their peers and construct a rough draft of a graph to show their predictions;
- measure the distance of their classmates' jumps;
- describe and summarize the shape of the jumping data;
- generate ways of identifying a typical jump;
- organize jumping data on a line plot.

Prior Knowledge

Students should have previous experience with measuring the distance between two points in inches or centimeters. Be sure that they know how to use a ruler or meterstick properly.

Materials and Equipment

- Graph paper
- Several rolls of adding-machine tape or thick string or yarn
- Rulers or measuring tapes
- Meterstick
- Mathematics journals or notebook paper
- A transparency of the blackline master "Summer Olympics 2000"

p. 101

Classroom Environment

- Students work in pairs on the activity.

Some ideas for this activity were adapted from Economopoulos et al. (1995).

Information about the history of the long jump event can be found by searching on "long jump" at the Britannica Web site (www.britannica.com).

Fig. **2.1.**
Results of twelve finalists

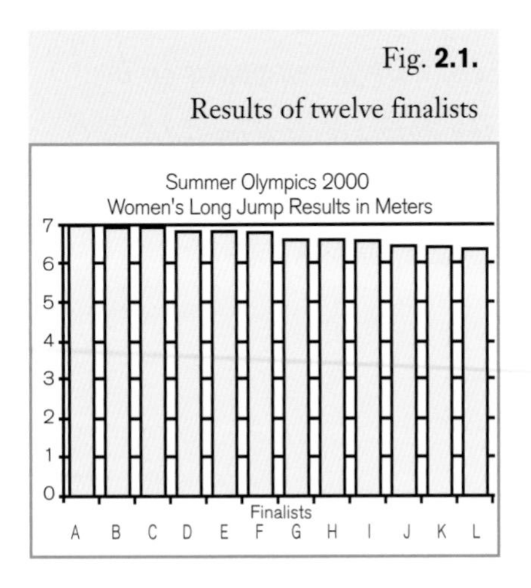

Students can read about Marion Jones on the USA Track and Field Web site (www.usatf.org/athletes/bios/Jones_Marion.shtml).

Activity

Engage

Introduce this activity by sharing some information about the long jump event in the Olympics. Tell students the following: "This event was once performed in two different ways, with both a running and a standing start. The long jump from a standing start was discontinued after the 1912 Olympic games. With the running start, long jump athletes begin on a runway that is about 150 feet long. They try to build up speed and jump as close to the end of the runway as possible and land in a sand-filled area. In the 2000 Summer Olympics in Sydney, Australia, a woman from the United States named Marion Jones came in third in this event, winning a bronze medal."

Marion Jones's jump was just a little less than 7 meters, or about 22 feet, long. Show students the length of 1 meter with a meterstick and ask students to estimate how long her jump was. Together, find a length in the classroom that is about as long as Marion Jones's jump. You may want to cut a piece of adding-machine tape that is a little less than 7 meters long (6.92 m), since students will use adding-machine tape later in the investigation to show the length of their jumps.

Next, show students a graph (fig. 2.1; blackline master "Summer Olympics 2000") of the results of the twelve finalists in the long jump in the 2000 Summer Olympics. Ask them to find the bar on the graph that represents Jones's jump (bar C). Engage students in a discussion about this bar graph by posing questions such as "What information does this graph give us?" "How is it organized?" (It is a bar graph. Each bar shows us the length of the jump of one finalist. The bars are ordered from longest to shortest.) "What statements can you make about the data on this graph?" (The twelve athletes' jumps were all between 6 and 7 meters. The top three jump lengths were very close. Most of the jump lengths were more than 6 1/2 meters.) "How does Jones's jump compare with those of her competitors?" (She jumped almost as far as the second-place finisher.)

Next, move the discussion about the data toward considering what might be a typical jump for these top athletes. Ask students, "From these data, what do you think is a typical jump for a female Olympic athlete?" Students at this point may say that it looks like the typical length is about 6 1/2 meters because several of the jumps are near 6 1/2. Others might offer a range, saying the typical length is greater than 6 1/2 and less than 7 meters. Ask students to predict what a graph of the results from the next Olympic long jump finals might look like.

Explore

Predicting jumping performance. Tell students that you will be asking them to predict once more, but this time they will imagine that twelve students their age have participated in a long jump from a *standing* start. Direct students to work with a partner: "With your partner, make some predictions about the jump lengths of these twelve students. Decide what you think a bar graph of the data might look like. Draw a sketch of the graph on graph paper. Underneath your graph, write and explain two or three predictions that you have about how the data might look."

As students work, observe them and note how they approach setting up bar graphs. Students sometimes have difficulty deciding what to put on each axis. Help them by referring to the bar graph of women's long jump results that they examined earlier. They sometimes start the numbers on the y-axis at the smallest number in their range. Show them that the y-axis should start at 0. Be sure that they know that the increments on the y-axis should be equal and that there must be one bar for each of the twelve jumpers.

Ask students to post their graphs on the chalkboard or a bulletin board for all to see, and give the student pairs enough time to examine their classmates' graphs. Ask them to identify (*a*) predictions that are similar to or different from those they made and (*b*) graphs and predictions about which they have questions. Ask them to record the similarities, differences, and questions in their mathematics journals or on an index card so that they can refer to them in the following discussion.

With the students all together, discuss how they made their predictions. Encourage them to ask questions about one another's graphs and predictions. Ask questions such as "What distance do you predict the shortest and longest jumps might be?" "What do you think might be the length of a typical jump?" "Do you think the jumps will mostly be the same?" "Why?" Record students' predictions on chart paper or on the chalkboard to refer to after they have collected actual data on how far students can jump.

In data analysis the word *range* is used to indicate the difference between the maximum and minimum values. For example, if data on the lengths of the jumps of twelve students are collected and the shortest is 1 foot and the longest is 4 feet, the range of these data is 3 feet.

Collecting jump data. Explain to students that they will now collect data to test their predictions. Instead of just thinking about twelve imaginary students, the class will collect actual data on how far each student in their class can jump.

Describe the task to students: "Working with your partner, you will take turns jumping and measuring the lengths of your jumps. Put a piece of tape on the floor as a starting line. When you jump, start with your feet together and your toes touching the line. After you jump, stay where you landed. Your partner will use a piece of adding-machine tape to measure the length of your jump from the beginning of the starting line to the heel of the foot that is closest to the line." You may want to have one pair of students help you model how to measure so that all students' jumping and measuring methods are consistent. Decide whether or not students will be allowed to take a practice jump. Maybe all students will jump two or three times and record their best jump. Remind students that, unlike Marion Jones's jump, their "long jumps" will be two-footed jumps starting from a standing position; that is, the students will start on two feet and land on two feet.

When students have completed their jumps, have them measure the lengths of their adding-machine tapes and then post them on the chalkboard. You may need to have them put a large paper clip on the end of the tapes to keep them from curling as they hang on display.

Arranging the tapes of the jumps from shortest to longest creates a bar graph of a type known as *case-value plot.* In a case-value plot, the height of each bar represents the value for a single data element–that is, for a single case.

Analyzing the jump data: How can we find the typical jump length? Bring the class together to examine the data. Suggest to students that organizing the pieces of adding-machine tape may help them to get a better picture of the data. Ask students how they could organize them, and if no one suggests it, have students organize the tapes from shortest to longest. This will create a bar graph. Guide students in their examination of the data by asking these questions: "What is the range of the data?" "Are there any unusually large or small jumps?"

Next, ask students to consider how they could use all the data to find the length of a typical jump for students in their class. Ask each pair to join another pair and discuss this idea together. Allow time for discussion and then have students write in their mathematics journals or their notebooks about how they would use the data to find the typical jump length for students their age. Note that some students may want to select one value as the middle-sized jump whereas others may want to select a range of values. Have students share their ideas with the rest of the class. Allow time for discussion.

Extend

Making a line plot. On a large piece of chart paper or on the chalkboard, model the process of making a line plot (see fig. 2.2). Tell students that one way to analyze their jump data is to organize them in the form of a line plot. Think aloud about the process of setting up the line plot and recording some values:

> A line plot is a type of graph that we can make quickly. It will help us to see the range of the jump lengths just as our arrangement of the adding-machine tapes did. It will also help us see other important features. We will be able to see how the jump lengths are spread out. Just by looking at a line plot we can see if there are any unusually long or short jumps. We can see if there are several jumps that are close to the same length. To make a line plot, we first need to determine the range of the data. What is the least value? What is the greatest? To start making the line plot, we draw a horizontal line. Below the line we write the smallest number, or least value, all the way over on the left side of the line, and then we write the numbers between the least and greatest values until we reach the largest data value, on the right side. Then, for each person, we write an X above the number that tells how long her or his jump was. It's important to make the Xs about the same size.

Making the line plot on graph paper or chart paper, with one X in each cell, can help students keep all the Xs the same size.

Fig. **2.2.**

Sample line plot of jump lengths

				X					
			X		X				
	X		X	X	X				
X	X		X	X	X				X
96	97	98	99	100	101	102	103	104	105

Jump Length (centimeters)

This is a sample line plot. Note its simple setup. Line plots can be made quickly and can reveal important features of a data set.

Model plotting a few of the jumps on the line plot. You can continue to plot the rest of the values, or you can have students come up to plot their own values.

Engage your students in a discussion about the different aspects of the data shown by the two graphs—the bar graph made with the adding-machine tape and the line plot. Unlike a bar graph, such as the one made with the adding-machine tape, a line plot shows the values for which there are no occurrences of data. It allows us to see how the data are distributed, including where there are clusters of data, where there are holes in the data, and where there are any unusual values in the data set.

If this investigation is the first time your students have made line plots, don't assume that they will be able to make them independently during your next data investigation after seeing you model it one time. Students often need more experience constructing these graphs. One common mistake that students make when they begin to use line plots is not writing *all* the values between the least and greatest values. Sometimes students will write the least and greatest values but in between write only the data values that are in their data collection. For example, if they were making a line plot with the numbers 20, 20, 23, 25, 25, 25, 26, 29, students might mistakenly make a line plot leaving out 21, 22, 24, 27, and 28 below the horizontal line. If you see your students doing this, help them construct the line plot correctly by discussing the effect of leaving out those values. Give them some guided practice using small data sets. Some students will need additional modeling. Using line plots

"Compare different representations of the same data and evaluate how well each representation shows important aspects of the data."

(NCTM 2000, p. 176)

frequently during informal data collection throughout the year is one way to help students understand how to set up these very useful tools for data analysis. Guide students in analyzing the jumping data represented on the line plot. Encourage them to make new statements about the class data. Ask such questions as—

- "What do you notice about our jumping data?"
- "Are there any clusters of data—that is, places where there are many Xs?"
- "Why do you think these clusters exist?"

(The clusters show that there are groups of students who jumped about the same distance. Students may speculate that there are clusters because the jumpers are all about the same age or about the same height.) Continue probing by asking such questions as—

- "Are there any gaps—that is, values on the line plot where there are no Xs?"
- "Are there any unusual jump lengths?"
- "How can we explain why these unusual jump lengths exist?"

(Students might suggest that these unusual values point out that some students can jump a lot farther than others. They might also suggest that these unusual values could be due to an error in measurement or recording.) Pursue the ideas by asking—

- "Why do you think our jumping data look the way they do?"
- "If we repeated the process of collecting jumping data tomorrow, what do you think the line plot would look like?"
- "If we collected jumping data from students in first grade or in sixth grade, what do you think the line plots would look like?"

As you discuss the features of the data, students will probably point to the value for which there are the most data. Explain to students that this value is called the *mode*. Ask them to record three statements they can make about the jumping performance of the class from the line plot. Encourage them to focus on how the data are spread out and what they notice about the shape of the data.

The investigations in this chapter introduce students to important terms in data analysis, such as *mode, median, mean, average, outlier, range,* and *distribution.* Teachers introduce these words to students in the context of data collections, especially while discussing students' observations of the data. You may need to offer some additional support to students whose facility with the English language is developing. Be sure to look for ways to use the terms in subsequent discussions with students.

Assessment Ideas

When assessing students' understanding in this investigation, pay particular attention to the statements students write or say when you ask them to describe or summarize the data results. You want students to move beyond identifying their own data values in the collection and noting which values are the greatest or the least. You want them to examine the collection of data as a whole and consider the distribution of the data. This takes time and is difficult for students who are just beginning to analyze data.

Where to Go Next in Instruction?

If this investigation is one of your students' first experiences with collecting data, organizing data on a line plot, describing and summarizing a data set, and looking at ways to find what the "typical value" is, you may want to follow up the investigation with additional data collections. There are numerous possibilities for collecting data. Students are particularly fascinated and motivated when collecting data on themselves and their peers. Some suggested questions for students to investigate with data include the following: How many siblings do you have? How many pets do you have? How many hours of TV do you watch each week? How many books do you read each month? How long is your little finger? How long is your foot? How many jumping jacks can you do in 30 seconds? How many marbles can you pick up in one hand?

After conducting one or more of these types of data collections, ask students to represent the data collected on a bar graph, tally chart, or another type of graph or chart. Ask them to compare the type of information they are able to see when examining the line plot of the data with the type of information revealed about the data in other representations.

This investigation can also be extended by collecting additional data on jumping. Students may be interested in comparing their jumping data with those of another class of the same grade or with a classroom of younger or older students. Students can be asked to collect these data, construct line plots, and look for the typical jump length in each group. They should be asked to compare the results of these data collections with their class's jumping data and to come up with reasons for similarities and differences.

Students can also be asked to jump a second or third time and add these data to the line plot. Examining how the shape of the data changes with the addition of new data values can lead to many interesting discussions.

This investigation involved students in collecting data and then describing the shape and features of data sets. It introduced learning how to summarize data by considering the typical value in the data set. The next activity, How Many Stars Can You Draw in One Minute? continues to build on this idea while presenting students with another opportunity to use line plots as organizing, recording, and analysis tools. In this investigation, students begin to make comparisons of data sets by comparing the shape of the data on line plots.

How Many Stars Can You Draw in One Minute?

Grades 3–4

This activity develops students' ability to describe and summarize the distribution, or shape, of a data set. Students summarize data by indicating what is typical of the data. They also begin to make comparisons among data sets. In this investigation, students collect data on how many stars they can draw in one minute. They examine a line plot of another class's data collection in the same activity. They summarize and describe the features of their line plots and decide on a typical value in the data. Students also compare line plots, examining the shape of the data, and generate hypotheses to explain the differences.

"Describe the shape and important features of a set of data and compare related data sets, with an emphasis on how the data are distributed." (NCTM 2000, p. 176)

Goals

Students will—

- collect and represent data on line plots;
- describe the shape of data;
- summarize the data, indicating what is typical of the data;
- compare two sets of data on line plots.

Prior Knowledge

Students should have previous experiences in making line plots.

Materials and Equipment

- A class clock with a second hand visible to all students
- Sticky notes
- A copy of the blackline master "How Long Is One Minute?" for each student
- A transparency of the blackline master "How Many Stars?—Another Class" and paper copies for each student
- Mathematics journals or notebook paper

pp. 102, 103

Classroom Environment

Students work in pairs and in groups of four during this activity.

Activity

Engage

Thinking about one minute. Engage students in a conversation about how long one minute is. Depending on their previous experiences with time, students may not have a good sense of the length of one minute. Ask students to brainstorm tasks they could do in one minute. Give your students some experience in thinking about one minute by asking them to find the number of times in one minute they can do the following: snap their fingers, bounce a basketball, write the word *math*, count by fours or fives, hop on one foot, and so on. Have students predict how many times they will be able to do each task. Students should record their predictions and then their actual results on copies of the blackline master "How Long Is One Minute?"

While conducting these one-minute tasks, take some time to talk to your students about counting strategies. Counting by ones isn't a very efficient strategy for these tasks. Ask students to suggest ways to count more efficiently. If no one suggests it, propose that for writing the word *math*, students could circle groups of ten *math*s, or for counting by fives, they could reason about how many fives they had counted. They could think, for example, "If I stop at 85, 80 has twice as many fives as 40. I know there are 8 fives in 40, so there are 16 fives in 80. Eighty-five has one more group of five, so I counted by fives 17 times."

Bring students together to share their results from the one-minute tasks. Go over the predictions and results that the students wrote on the activity page "How Long Is One Minute?" Select two of the tasks and have students write their individual results on a class results chart that you create (see fig. 2.3). Direct class discussion by posing the following questions: "What do you notice about the class results?" "Were you surprised by any of your results or by any of the class results?" "Can you think of other tasks that you could do many times in one minute?" Students probably know that a minute is sixty seconds, but they may be surprised by the actual magnitude of one minute or how many times they can perform a task in one minute.

Fig. **2.3.**

Class results chart to show tallies for one-minute tasks

How Much Can I Do in One Minute?—Class Results				
Snap Fingers	Bounce Ball	Write *Math*	Count by Fives	Hop on One Foot

Later in this activity, students will be using a line plot to record data. They will also be looking at line plots with data from other groups of people. Depending on your students' previous experience in making line plots, you may want to have them make a quick line plot for one of the tasks they have just performed before moving on. For example, have them make a line plot showing the number of times students counted by fives. Ask the students questions: "What values do we need to list on our line plot below the horizontal line?" "What is the range of our data?" "What are the smallest value and the largest value?" Have students come up to the chalkboard or chart where you are making the line plot and put an X above the number of times they were able to count by fives. Next, spend some time having students summarize and describe any data patterns or trends that they see. When you ask students to describe the data, they will probably point to the value that occurred the most. Remind students that this is called the *mode*. Then encourage them to examine the overall features of the distribution of the data. Ask them questions such as "How are the data spread out on the line plot?" "Are there any data clustered together?" "What do those clusters tell us?" "Are there any gaps?" "Does the shape of the data surprise you?" "Are there any unusual values?" "Why do you think our data have the shape they do?"

Explore

Tell students that a class of twenty students in another school collected data on how many stars each student could draw in one minute and that you have a line plot showing their results. Before showing them the line plot on the blackline master "How Many Stars?—Another Class," ask students to predict what they think the range of the data values on the line plot will be and what they think the shape of the line plot might look like. Ask them to sketch their predicted line plots in their mathematics journals and to explain their reasons for their predictions. Have partners share their work with each other before asking a few students to share their predictions and reasoning with the whole class.

Next, display a transparency of the blackline master "How Many Stars?—Another Class" on the overhead projecter and distribute a paper copy to each student. Have partners or groups work together to examine the line plot and discuss the questions on the activity page. They should complete question 1 together and question 2 independently. Bring the class together to share a few of the observations that students have made, particularly their statements about the data from the class in question 1. Begin the discussion with questions such as these:

- "What statements did your group record?"
- "What can you tell about this class's star-drawing performance?"
- "What would you say is the typical number of stars a student in this class could draw? Why?"

Students' responses to question 1 may be similar to the following: "There is a fairly wide range." "The smallest number of stars was 34, and the largest was 46 stars." "There is a cluster of data around 40–43." "Most of the students in the class drew between 40 and 43 stars." "Of the 20 students in the class, more than half of them drew between 40 and 43 stars." "A typical student in this class would probably draw about 41 or 42 stars." Next, have students share their predictions about the number of stars that they will draw in one minute (question 2).

Tell students that before they collect data on their class's star drawing, they need to establish some rules. Decide with students what kind of star each student should draw. Tell students that they will do some practice drawing before they begin the data collection. Have students complete question 3 and do some additional practice for homework.

Collecting star data. Distribute some blank paper to each student as well as a sticky note on which to record the results. Lead the star drawing by instructing students to begin when you say "go" and finish when you say "stop." When counting, students should group their stars in clusters of ten (or use another efficient counting method) and record their result on a large line plot using the sticky notes. Have students make their own copy of the line plot in their mathematics journals or on notebook paper. Have students double-check that their recorded line plots accurately correspond to the data collected.

How Many Stars?–Another Class

								X				
								X				
						X		X				
						X		X	X			
						X	X	X	X			
X			X		X	X	X	X	X	X		X
34	35	36	37	38	39	40	41	42	43	44	45	46

Next, ask students to summarize and describe the distribution of the data. Again, focus their attention on the shape of the data and the features of the entire data set. Use these questions to guide the discussion: "Are there any clumps or clusters of data?" "What might be some reasons for these clumps or clusters?" "Are there any gaps?" "Are there any unusually large or small data values?" Explain to students that these unusual data are called *outliers*. Outliers are either much higher or much lower than most of the other data values. Can they think of possible explanations for these outliers? What do they think is the typical number of stars drawn by students?

Ask students to work in groups to write a description of the data. They should include their group's decision about how to describe the typical number of stars that a student can draw in one minute. Tell students to be sure to give an explanation for their decision. When you bring all the groups together and ask each one to share its descriptions, encourage the other groups to participate in the discussion by asking questions of their classmates. When students share their thinking about the typical number of stars that can be drawn by students in their class, they may pick a range of numbers, the middle number, or a data value for which there is a high frequency of values.

If students suggest taking the number in the middle of the distribution, explain that mathematicians call the value in the middle of an ordered data set the *median* and that this number is one way to look at what is a typical value in a data set. The median is one type of average or measure of central tendency. It is the midpoint in a data set when the data values are put in numerical order. For example, in the following data set, 45 is the median: 33, 37, 39, 40, 40, 45, 46, 46, 46, 48, 50. When there is an odd number of data values, the median can be found in the data set. When there is an even number of data values, the median is the number in the middle of the two middle numbers. For instance, in the data set 33, 37, 39, 40, 40, 45, there is no one data entry that is exactly in the middle. The median in this instance is the number between the third and the fourth data values, 39 and 40. The median is 39.5. (The activity Do You Get Enough Sleep? develops students' understanding of median.)

Extend

It is important for students to have considerable experience comparing related data sets. By comparing sets of data, students can learn to focus on describing and summarizing the entire data set; they can look for patterns and trends and generate hypotheses to explain them. Have students look back at the line plot that you showed them earlier in this investigation on the blackline master "How Many Stars?—Another Class." Bring students together to share some of their responses after you have asked them to write in their mathematics journals about questions like the following: "What do you notice about the shape of the data of that class's line plot compared with that of the data of our line plot?" "What might be some reasons for what you see?"

Assessment Ideas

When looking at students' work, see if they are making a transition from (*a*) descriptions about data sets that focus on discrete elements of the data, such as "Three students drew 40 stars" or "The largest number of stars drawn was 60," to (*b*) statements that consider the data as a whole or identify features of the data, such as "There is a large cluster of data between 50 and 60" or "More than half the students drew at least 50 stars." Also look to see if students are able to hypothesize about the shape of the distribution of the data—that is, if they are able to come up with explanations for why the distribution of the data has the shape it does.

Where to Go Next in Instruction?

You can continue making comparisons with related sets of data in many ways. Try investigating the following questions and comparing the data results with your class's original star-drawing data: "How do our data compare with those of an older or younger class?" "How many times can you draw a star with your nonwriting hand?" "Does practice make a difference?" Collect the data from students first without letting them practice and then a second time after allowing them to practice.

For homework, have students ask a parent or another adult to draw as many stars as he or she can in one minute. In class the next day, ask the students to compile their data to answer the question "Do adults draw stars faster than we do?"

After students have developed an understanding of how to describe the shape of a data set and can come up with ways of identifying the typical value in a distribution, they are ready to begin to explore the concept of median more closely. An important element of their learning about methods for analyzing data is learning about the three types of averages: mode, median, and mean. Each of these types of averages is easy to find, but you need to go beyond teaching students the *procedures* for finding them. For these statistics to be truly useful, students need to have a strong conceptual understanding of each and need to know how to interpret each within the context of an entire data set. Having students summarize data sets and identify what they think is typical is an important start. The activities that follow are approaches to developing students' conceptual understanding of median and mean.

Do You Get Enough Sleep?

Grades 4–5

Building on students' understanding of what is typical, this investigation introduces students to the concept of median. Students collect and organize data on the number of hours each student slept one night. Students decide on a typical amount of sleep for students in their class. Then they find and interpret the median. Students collect additional sleep data from other classes and make comparisons.

"Describe the shape and important features of a set of data and compare related data sets, with an emphasis on how the data are distributed; use measures of center, focusing on the median, and understand what each does and does not indicate about the data set." *(NCTM 2000, p. 176)*

Goals

Students will—

- identify the median of a data set and of data presented on a line plot;
- use the median to describe a set of data;
- use the medians of two sets of data to compare the distributions.

Prior Knowledge

Students should have previous experiences in collecting and organizing data on line plots. They should be able to round numbers of hours and minutes to the nearest quarter hour.

Materials and Equipment

p. 104

- One copy of the blackline master "How Much Sleep Do You Get?" for each student
- A transparency of the blackline master "How Much Sleep Do Children Typically Get?"

Classroom Environment

Students work in pairs and in groups of four during this activity.

Activity

Engage

For homework on the day before you do this activity, students should record the time when they go to bed and the time when they awake. On the next day, begin the lesson by distributing the blackline master "How Much Sleep Do You Get?" and having each student find the total time he or she slept the previous night. Students may ask if they should count from the time when they actually went to bed or the time when they fell asleep. As a class, decide what you will consider to be "sleeping time." Have students round their times to the nearest quarter hour. For example, if a student slept 8 hours and 27 minutes, she or he would round this up to 8 1/2 hours; if a student slept 8 hours and 12 minutes, she or he should round this to 8 1/4 hours. Ask them to reflect on whether they think the amount of sleep that they calculated is typical of the amount of sleep they normally get and to predict how their hours of sleep will compare with those of their classmates.

Have students share their work with a partner before you ask several of them to share their responses to questions 2 and 3. Ask students to consider whether or not they think they usually get enough sleep and if they think children their age generally get the sleep they need. Ask students to explain their thinking.

Show a transparency of the blackline master "How Much Sleep Do Children Typically Get?" or put the data from that sheet on a chart for all students to see. Tell students, "This chart shows information about how much sleep children in the United States typically get and how much sleep they need. What do you notice about the data?" Students may point out that according to the data, children are getting less sleep than they need, and that as children get older, they need less sleep. You might also ask students to consider the difference between the number of hours children typically sleep each night and the number of hours recommended. For example, the average number of hours a ten-year-old sleeps is 9 hours, which is 3/4 of an hour less than what he or she needs. Students may notice that the difference between the amount of sleep that children need and the amount they actually get grows as children get older. Ask students to hypothesize about why this is true. Ask them to consider the number of hours of sleep they think they usually get and compare it with what is typical as well as with what is necessary for children their age.

Ask students to make a prediction: "What do you think we will determine to be the typical amount of sleep that students in our class get? How do you think our class will measure up if we compare how much sleep the students in our class typically get with what is recommended?" Students will probably use their own amount of sleep and their own experiences when they explain what they think is a typical amount of sleep for their classmates. Tell students that they are going to organize the class's sleep data and examine this information to decide what is a typical amount of sleep.

Lead the class in constructing a line plot. The numbers below the horizontal line should increase by quarter-hour segments, starting at the smallest amount of sleep that any student recorded and proceeding to the largest amount. As you construct the line plot on the chalkboard or overhead projector, have students construct one in their mathematics journals. Have all students report their sleep times, and place the data on the line plot.

Next, ask students to examine the line plot. Have them work with a partner to describe the features of the data distribution. Ask students—

- "What can you say about the number of hours you sleep? Is it like most of your classmates?"
- "What do you notice about the shape of the data?"
- "How are the data spread out? Are there any clusters or gaps? Are there any outliers (any unusually large or small values)?"
- "What is the mode of the data? How can you use the data on the line plot to find a typical number of hours of sleep for students in our class?"
- "What are reasons that explain the shape of the data distribution?"

Post these questions on chart paper for students to refer to as they examine the line plot. Have them record in their mathematics journals at least four statements that they can make about the sleep data. Ask students to work with their partners to decide on a value or range of values that they think is typical of the number of hours of sleep. Direct students to record their decisions and their reasoning below the statements.

Explore

Introducing median. Provide students with an opportunity to share their decisions about what they think is a typical number of hours of sleep. Have each pair of students share its ideas with another pair before leading a whole-class discussion. Some students will point to the mode as what is typical, others will select a range within a large cluster of data, and some students may indicate that the number in the middle is typical.

Explain the term *median* to students: "Statisticians (persons who deal with data or numerical quantities) sometimes use something called the *median* to describe what is typical of a data set. The median is the midpoint of a set of data when the data are arranged in order. The word *median* actually means 'middle.' We could look for the middle number in our sleep data. The median is helpful because it lets us know that exactly half of the data values are above it and exactly half the data values are below it."

Before finding the median of the sleep data, demonstrate how to locate the median by asking seven volunteers to arrange themselves in a line in front of the class in order of height. Point out to students that the height of the person in the middle is the median height. You may want to have one person and then another and another move away from each end of the line to show how to find the middle value, which in this example is the fourth number. Have these seven people reform the line and have an eighth person join them, again lining up in order by height. Ask students how they could find the median now. If no one suggests it, explain that when there is an even number of entries, there are two "middle" values. The median in these instances is found by determining the average of the two values. When there is an even number of data values, the median is not one of the elements in the data set.

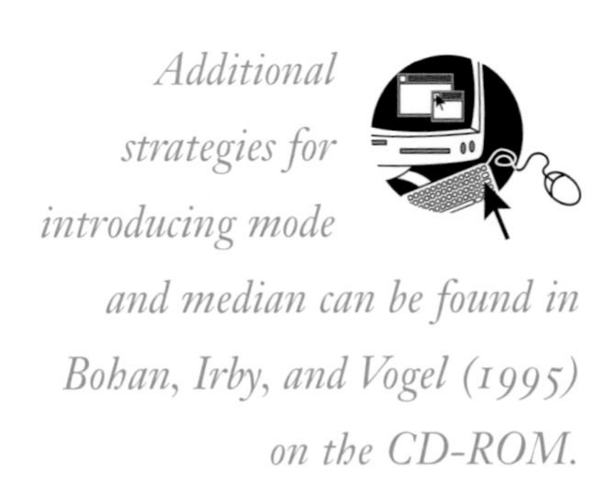

Additional strategies for introducing mode and median can be found in Bohan, Irby, and Vogel (1995) on the CD-ROM.

Ask students to work with their partners to locate the median of the sleep data and then to compare the median with what they thought was the typical amount of sleep. Next, ask them to compare the median of the class data with the average number of hours of sleep students their age get and the number of hours of sleep recommended.

Extend

An important next step in this investigation is to have students compare their class's sleep data with data from another class. Have students collect data in other classrooms. For example, one group of students can collect sleep data in a classroom with younger children, another group can collect sleep data on older children, and another group can collect sleep data on a group of adults. Have students construct line plots and examine the shape of the data for each age group and compare them with the recommended number of hours of sleep. Ask students to record similarities and differences that they notice among the data sets, as well as their hypotheses about why the data sets look the way they do. Have students write up their comparisons and recommendations to share with the classes from which they collected data. Ask them to write a letter to parents or other adults about their observations on the amount of sleep students are getting. They could also compare their class's hours of sleep on weekend nights with those on weekday (school) nights.

Another way to extend this investigation is to have students work in groups of three or four to collect other sorts of data on two groups of people and make comparisons. For example, students could compare the typical height of students in their class to that of students in kindergarten, determine the typical number of hours a week that girls watch TV as compared to the typical number of hours that boys watch, or find the typical number of fifth-grade students who order school lunch compared to the typical number of first-grade students who do so. Students enjoy creating a research question and collecting data about their peers. Be sure that the data that they collect will allow them to identify the typical *number* of something so that they will continue to explore what they learned about median (i.e., students cannot find the median favorite color or food; the median is not meaningful in the analysis of categorical data, which are not ordered numerically).

Assessment Ideas

As students collect additional sleep data, look closely at how well they are able to organize the data on a line plot, describe the features of the shape of the data, and identify a typical value. Determine if they are able to identify the median of the new set of data and interpret the median within the context of the rest of the data. Collect students' comparisons and examine how well they made comparisons between the two data sets.

Where to Go Next in Instruction?

In this investigation, students learned about median. Keep in mind that once students know how to determine the median, their thinking about a data set should not be reduced to just finding this statistic. The median of a set of data is a useful statistic. It is important that students understand that the median is helpful because it can give us an indication of what is "usual," "typical," or "average." However, they need to continue to use the median as one piece of information, and they also need to continue to examine the features of the entire data set. It can be helpful to students when looking at a line plot to learn how to "eyeball" the median: Ask them to look at the entire set of data on the line plot and imagine about where they could draw a vertical line so that half the data are on one side of the line and half are on the other.

Engage your students in a discussion about what information the median tells us about a set of data and what it does not tell us. For example, when we know the median of a set of values, we know that exactly half of the data are above the median and exactly half are below. When we just know the median of a data set, we do not know the range of the data, we do not know if there are any unusual values, and we do not know what the shape of the data is. One way to help students to consider this concept is to ask them to construct data sets when given a median. For example, ask students, "If the median number of siblings in our class is 3, what could the line plot of the data

See Zawojewski and Shaughnessy (2000) for additional discussion of students' understanding of median.

look like? Can you find more than one possibility?" Provide your students with additional opportunities to construct data sets for a given median.

There are some common misconceptions that students have as they begin to learn about and use the median. Look for these as you continue instruction. One misconception is that the median is the middle number in the range. What is really the same error commonly happens when students are finding the median from data organized on a line plot. They will pick the middle value on the horizontal axis instead of the middle data entry. To avoid reinforcing these misconceptions, provide your students with opportunities to find the median of sets of data both when the data are presented on a line plot and when the data are given as a collection of numbers. Present students with sets of data in which the median is close to either end of the range and engage them in discussion about this type of distribution.

Exploring the Mean

Grade 5

There are three types of averages in statistics: the mode, the median, and the mean. When the term *average* is used, it is often interpreted to be the arithmetic mean. Although students can understand the procedural definition of mean and compute it by finding the sum of the data values and dividing this sum by the number of data values, the concept of mean is difficult for students to understand. Conceptually, the mean can be thought of as a balance point in the distribution. Students need experiences like the one described in the following activity to understand this. It is also difficult for students to understand the relationship between the mean value and the set of data it represents.

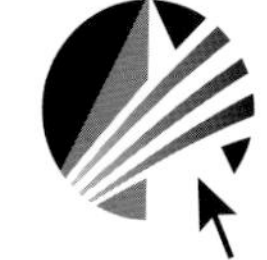

"Use measures of center, focusing on the median, and understand what each does and does not indicate about the data set."

(NCTM 2000, p. 176)

Goals

Students will—

- identify the three types of averages: mode, median, and mean;
- describe the mean as a balance point in a data distribution;
- create data sets for a given mean;
- estimate the mean of small sets of data.

Prior Knowledge

Students should have previous experiences with finding and interpreting the median and mode of a data set. They should have already learned how to summarize and describe the shape of a data set, including the range of the distribution.

Materials

- Snap cubes or Unifix cubes (14 for each group of 3 or 4 students)
- Chart paper or blank transparencies for recording two lists
- A sheet of unlined paper for each group of students
- Sticky notes
- A copy of the blackline master "Women's Soccer Results" for each student

p. 106

Classroom Environment

Students work in groups of three or four on the activity.

Activity

Engage

Begin by talking to students about the term *mean*: "You may hear the word *average* used a lot in real life. Although there are different types of averages, when people use the word *average*, they are usually referring to the *mean*, or arithmetic mean. The mean, like the median, is a value that can be used to summarize or represent a data set. It is another statistic that we can use to help us think about what the typical value of a data set is." Some of your students may have already learned how to find the

Some ideas for this activity were adapted from Friel, Mokros, and Russell (1992).

mean (adding the data values and dividing by the number of values). Explain to students that they will be investigating what the mean is and how it relates to the data.

Averages are often used in sports to summarize or indicate a team's or a player's performance. What sports averages do your students know about? They will probably report having heard about batting averages and free-throw averages. Ask them to consider what those averages tell them about a particular athlete's or team's performance. Students may say that they tell us how well a baseball player hits or how accurately a basketball player shoots free throws. Others may point out that the averages are sometimes used to compare athletes. Ask them to consider what we *don't* know if we know just such an average. This will be a harder question for students to consider. You may need to ask additional questions, such as "Does knowing a basketball player's free-throw average tell us how many times the player shot a free throw? Do we know the smallest number, or the largest, of shots that the player has made in a game?"

Information for this activity comes from the USA National Team Game Results at the Women's Soccer World Web site (www.womensoccer.com/refs/usteamres2.shtml).

Give the students an example involving the U.S. Women's Soccer team: "In the last seven games of the 2000 season, the U.S. Women's Soccer team had an average of 2 goals per game (a goal = 1 point). What does this number tell us about the way the team scored during those games? What can't we tell by knowing this statistic? Did the players score 2 goals in each of the seven games? Our experience from watching and playing in games tells us that it is more likely that in some of the seven games they scored fewer than 2 goals and in other games they scored more." On chart paper or overhead transparencies, make two lists with your students: "What We Know about the Team's Performance" and "What We Don't Know about the Team's Performance." (Students' responses may include the following: "We know that the average number of goals in seven games is 2; some games may have had more than 2 goals, and some may have had fewer." "We don't know the score for each game or whether any games had particularly low or high scores." "We know that the total number of goals in the seven games is 14.")

Explore

Have students make a paper mat on which to line up their cube trains:

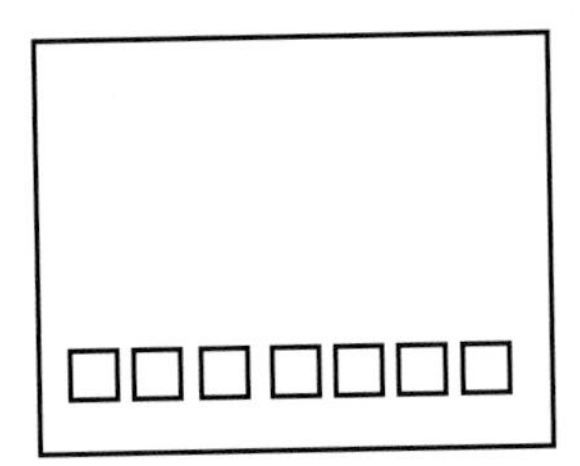

One pattern of rearranged trains is shown:

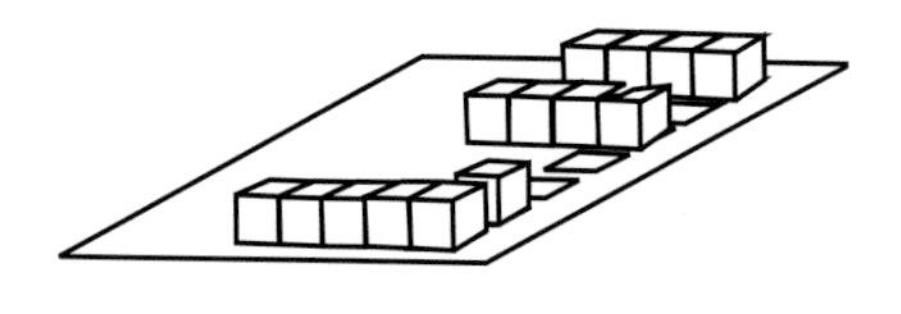

Show students seven cube trains with 2 cubes in each train. Tell them that the number of cubes in each train equals the average number of goals scored in seven games and that you have arranged them so that for each game there were 2 goals. Tell students that this is one possibility for the number of goals. Next, distribute a set of seven 2-cube trains to each group, and explain that the group members will work together to rearrange the cubes to make another set of data. Give each group a sheet of unlined paper, and ask the students to trace around one of the cubes to make a row of seven squares on which they will line up their trains. This mat will help them keep track of games with zero goals.

Encourage students to think about how they know that the data set they have created with the rearranged cubes could be a reasonable solution. Have students record their solutions and explanations in their mathematics journals. Bring students together and have each group share its solution and reasoning.

Next, use sticky notes to make a line plot on the chalkboard. Be sure that the numbers below the horizontal line range from 0 to 14. Each sticky note represents one of the games played. Arrange all seven sticky

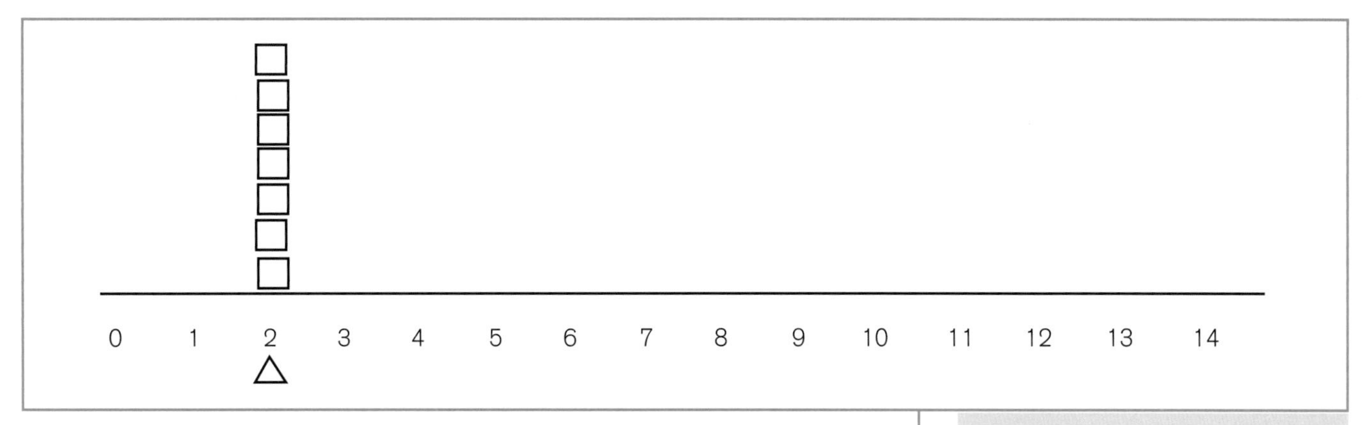

Fig. **2.4.**

Finding the mean number of goals in seven soccer games

notes so that each is above the 2. (See fig. 2.4—note that the triangle in the line plot indicates the mean.)

Explain the line plot to the students: "Instead of using cube trains to represent each game, we will use sticky notes on a line plot. This line plot shows one possible data set for the seven soccer games played by a team. The team scored 2 points in each game. The mean of this data distribution is 2. What is the median, or the middle value? (2) What is the mode? (2)." Tell students that for this particular data set, the three averages—the mode, the median, and the mean—are all the same number.

Explain to students that the mean can be represented as a point on a number line where the data on either side of the point are "balanced." Stress the idea that it is the distances of the data from the mean that must balance and not the number of data pieces on either side of the mean. To illustrate this idea, tell students: "Right now on this line plot there are no values above or below the mean). However, what would happen if we moved one of the sticky notes to a value above the mean?" Move one of the sticky notes to 3 and ask students to determine if the mean is still 2. (The mean has changed—there are six sticky notes representing 2 points each and one sticky note representing a score of 3 points. If we compute the mean for this data distribution, we find that the mean is no longer 2 but about 2.1.) Ask, "How can we change the distribution and keep the mean of 2?" If no one supplies an answer, suggest one: "Now let's move one sticky note below the mean of 2 to 1." (See fig. 2.5.) The distance from each of these two new points to the mean is the same (1); therefore, the mean should not change. (5×2 points + 1 point + 3 points = $14 \div 7 = 2$.) The mean is the same.

Pursue the idea: "How else can we change the distribution of the data so that the mean remains the same? How do you know the mean remains 2?"

Fig. **2.5.**

Finding the mean number of goals

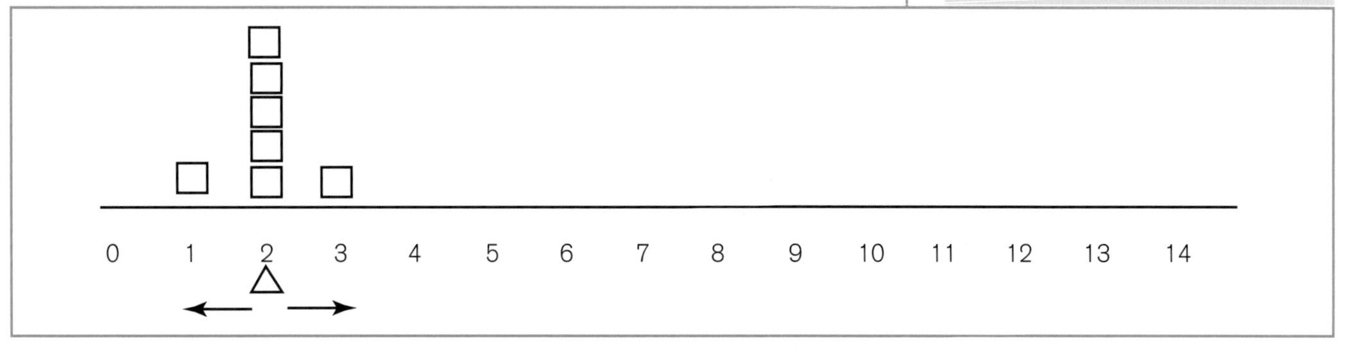

Suggest particular examples to develop students' thinking: "What would we need to do to keep the mean at 2 if we moved another sticky note from the 2 to 0 to indicate that one of the games had a score of 0?" (see fig. 2.6). Students should suggest that by moving one value 2 units below the mean, they need to move one of the data values to a place 2 units higher. Most students will suggest moving another sticky note from 2 to 4. Discuss how this will work, but also ask students to consider another suggestion. For example, the sticky note now at 3 could be moved to 5. You may want to use arrows to show that the sum of the differences above and below the mean are the same. Do some additional reconstructing of the data with students. Ask them the following question: "What if one game scored 5 goals?" "What if there were three games with no goals?" "What if no games scored less than 1 goal?"

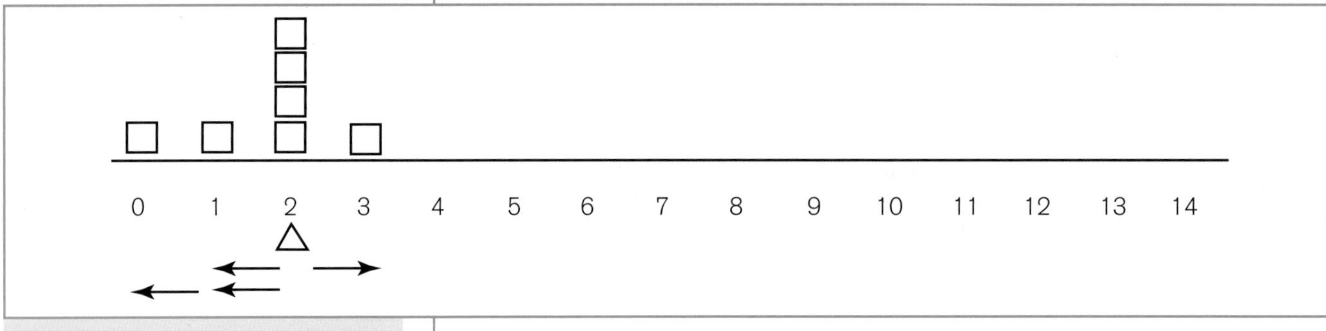

Fig. **2.6.**
Finding the mean number of goals

Extend

Distribute the blackline master "Women's Soccer Results" and ask students to work on it in their groups. Question 1 asks them to create two new line plots for the data values for the seven soccer games. As students create their data sets, you may need to suggest that they use some sticky notes or cubes to make their own movable data sets. (Some students will need to do this; others will prefer to work directly on paper.) Question 2 asks students to create additional sets of data with given parameters—that is, data sets in which (*a*) only one game had 2 goals, (*b*) exactly two games had 4 goals, (*c*) one game had an amazing 7 goals, and (*d*) the median of the data is 3. When students are first creating data sets for the given mean, they will probably make mostly symmetrical distributions, often by moving one value above the mean and then moving another the same amount below the mean. Others will see that they can move one value above the mean and then balance the distribution by moving more than one other value. The parameters in question 2 will present students with situations in which they will need to move many values. Question 3 asks them to write about the strategies that they used to create the data sets. Finally, question 4 asks students to create data sets for which the mean would be 4 instead of 2.

After students have had sufficient time to complete these questions, bring them together to share their results. Ask them to share how they know that their constructed data sets are reasonable. Be sure to pay special attention to distributions for questions 1 and 2 that do not have 2 as a data element. Ask questions like the following: "Did anyone create a set of data that has no games with 2 goals? How is this possible if 2 is the mean?" and "What did you need to do in order for one of the games to have a high number of goals, as in question 2(*c*) where one game had 7 goals?" Outliers (unusually high or low values) can affect the mean but

See Russell and Mokros (1996) on the CD-ROM for a useful summary of research on children's understanding of concepts of average, including mode, median, and mean, and a discussion of the difficulties that children experience and how teachers can develop their understanding of these concepts.

not the median. Extremely high data values or extremely low data values will increase or decrease the mean. In question 2(*c*) students would have to "balance" the high value of 7 by putting several values below 2.

Assessment Ideas

As you listen to students' responses and read their work, look to see that they are able to manipulate data on the line plot so that the sum of the distances on one side of the mean balances the sum of those on the other side. Notice how they are approaching the problems in which there are some parameters. Consider the following questions as you observe students: "Are they able to identify how a large value affects the distribution?" "Do they understand that 2 does not need to be one of the data values in order for the mean to be 2?" "Are they able to explain their strategies for constructing the data sets?"

Where to Go Next in Instruction?

In this investigation, students are introduced to the idea of mean as a balance point in the distribution and then explore this concept by constructing sets of data for a given mean. Provide students with some additional problems in which they are given the mean and need to construct a set of data. For example, ask students to create prices for six music CDs so that the average price is $9.00. Ask them to generate more than one solution, including one that does not have any CDs with a price of $9.00. The computer software Graph Master (Tom Snyder Productions 2001) can be used to provide students with additional experiences with mean.

Next, have students find the mean of a data set. Do not rush to teach them the procedure for finding the mean at this point. It will be important that students spend time estimating the mean and finding the mean by considering how the mean acts as a balance point in a distribution. For example, have students collect data on the number of siblings that ten or twelve students have and organize those data on a line plot. Be sure to have students describe the shape of the data distribution, and then have them estimate where the mean is and explain how they figured their estimates. Then ask students to find the exact mean. Do not encourage them to use the "add them up and divide" method. Instead, have them think back to the strategies they used with the soccer problem. After they have had time to find the mean, bring the students together to discuss strategies.

If the data set has some unusually low or high values, ask students to reflect on how those values affect the mean. Ask them to consider how the mean would change if there were no outliers. Ask students to consider what would happen to the mean if they collected additional data. Have them consider what happens to the mean if there are other changes in the data set. They will probably notice that if there is an increase or decrease in one value, the mean will also increase or decrease. Be sure to guide a discussion about what happens to the mean if one of the values is zero. Sometimes students will not count zero as a data value.

Continue instruction by having students find the means of other small data sets. Students could find the mean number of hours of television watched by students per week, the mean number of letters in the first names of people in their group, the mean height of a group of

See Mokros and Russell (1995) and Friel (1998) on the CD-ROM for additional information on research about children's conceptions of average, along with strategies for helping children build an understanding of mean.

students, or the mean number of books read in one month by a group of students. Once students have had experiences with mean, they can begin to include this statistic in their data analysis. They should start to consider this as another number that they can use to describe what is typical about a data set. As students conduct future data investigations, be sure that they consider the mean in the context of the distribution of the data and the other features of the data, including the median, the range, outliers, clusters of data, and gaps.

The investigations in this chapter are examples of ways to develop students' understanding of methods for analyzing sets of data, including how to analyze, summarize, and describe data. These activities have emphasized learning how to describe the shape and important features of a data set and how to use and interpret measures of center. The following chapter introduces approaches to helping students learn how to make and justify conclusions and predictions that are based on data.

Grades 3–5

Data Analysis and Probability

Chapter 3
Inferences and Predictions

"Propose and justify conclusions and predictions that are based on data and design studies to further investigate the conclusions or predictions."
(NCTM 2000, p.176)

We can gain a better understanding of the world around us when we are able to analyze data to develop and evaluate inferences and predictions. Collecting data, making inferences, and proposing predictions are all interconnected processes. To understand these ideas, consider the following:

- Data are collected through systematic means, such as observations, surveys, or experiments.
- Inferences are conclusions leading to a decision based on information contained in the data.
- Predictions are hypotheses about future events.

These concepts can be illustrated in the following exercise. Ask your students to conduct trials to gather some data to answer the following question: "What is the relationship between the drop height and the bounce height of a ball?" The students should make observations when they drop a ball from different heights, recording for each drop height the height of the first bounce. Then have the students make *inferences* about the relationship between the drop height and the bounce height. Ask them to *predict* the height for the first bounce of the ball when it is dropped from 100 centimeters. Have the students test their predictions by dropping the ball from 100 centimeters. If their estimates are not within a range of ± 7 centimeters, the students should gather additional data before making further predictions about the relationship. The systematic recording of data is essential in such a process. Students can use a table such as the following to organize the data collected. You might

When students explore the relationship between the drop height and the bounce height of a ball, they should–

- simply release the ball (not give it a downward thrust);
- let the ball bounce on a hard surface, such as a tile floor (not carpet);
- try other drop heights in addition to those suggested in the table; and
- repeat the experiment with different balls.

extend this activity to have the students determine whether the relationship holds true for other types of bouncing balls.

Bounce Height of a Dropped Ball

	Trial 1	Trial 2	Trial 3	Trial 4	Trial 5
Height of drop	40 cm	50 cm	60 cm	70 cm	100 cm
Measured height of first bounce	____ cm	____ cm	____ cm	____ cm	
Predicted height of first bounce for trial 5					____ cm
Measured height of first bounce for trial 5					____ cm

Through a variety of similar activities, this chapter addresses the development of concepts and skills appropriate for students in grades 3–5 related to the development and evaluation of data-based inferences and predictions. Students will learn to—

- evaluate predictions as a result of investigations;
- recognize how conclusions about the data might change with the collection of additional data;
- ask questions to make inferences such as "If this happens, that will follow" and "If I do this, that will happen";
- discuss possible conclusions supported by the data;
- distinguish between what the data show and what might account for the results;
- begin to recognize the cyclical nature of investigations (i.e., results often lead to continued problem posing);
- recognize that individual samples may vary;
- recognize that samples of the same size from the same population can yield different results.

One of the major goals of these activities is to help students to make predictions based on observations and evidence that they have gathered and to test their predictions for dependability.

The first activity, The Foot, the Whole Foot, and Nothing but the Foot! is designed to give students an opportunity to make predictions and inferences based on random samples from classroom and school data.

The Foot, the Whole Foot, and Nothing but the Foot!

Grades 4–5

This activity involves an exploration of the relationship between samples and the population from which they are drawn. A subset of the class (a sample) will collect and graph data about the length of their feet. Using the results of the sample, students will make predictions about the length of the feet of all the students in their class (the population). In this activity, students will also learn about random samples and how both methods of selection and the size of the sample affect the extent to which the sample can be considered representative of the population. Using a similar procedure, groups of students in the class will be asked to develop a plan for gathering data from a random sample of students in the school and to use the results to make a prediction about their school population.

Goals

Students will—

- collect and display data representing the foot length of a group of students in the class (the sample) to predict information about the whole class (the population);
- collect and display data about the foot length of a random sample of students in the school to predict information about the school population.

Prior Knowledge

Students should be able to measure length to the nearest centimeter and display data in a line plot.

Materials and Equipment

- Paper and centimeter rulers for each student
- One paper copy of the blackline master "The Foot, the Whole Foot, and Nothing but the Foot—Group Data" for each group, plus one transparency of it for use with the whole class
- One paper copy of the blackline master "The Foot, the Whole Foot, and Nothing but the Foot—Class Data" for each student, plus one transparency of it for use with the whole class

pp. 108, 109

"A population refers to all of the things, events, or individuals to be represented, while a sample refers to any number of things, events or individuals less than a population." (Christensen 2001, p. 49)

Classroom Environment

Students work in groups of five or six on the activity.

Activity

Engage

To do this sampling activity, direct the students to count off by fours, and let each student be identified by his or her number. Ask all the "ones" to form a group, and the "twos," "threes," and "fours" to do the same. Tell the groups that you want them to create a line plot representing the length of each member's foot to the nearest centimeter. Each student should record the length of his or her foot on the group's copy of the blackline master "The Foot, the Whole Foot, and Nothing but the Foot—Group Data."

Have the class consider various measurement options and agree on a common way to do the foot measurements. One method might involve removing the right shoe and stepping on a centimeter ruler to determine the length to the nearest centimeter. Another method would be to draw around the entire foot on paper and measure the length from the outline. Students may wish to predict their foot lengths before measuring their feet.

Select one group (sample) to record the resulting line plot on a transparency of the blackline master (or to draw it on the board). Engage the class in a discussion that addresses the following questions, and record any numerical data that are shared:

- "What does the line plot show?"
- "What is the smallest foot length in centimeters from this sample of the class population?"
- "What is the greatest foot length in centimeters from this sample of the class population?"
- "What is the range?"
- "What is the most common length among students in this sample of the class (the mode for this set of data)?"
- "What is the median foot length for this sample?"
- "Would you predict that this group's line plot is representative of a line plot of the entire class's foot lengths? Why, or why not?"

Explore

To do this phase of the activity, have the students predict what they expect the median and range of the total class's foot lengths to be, on the basis of their knowledge of the discussed sample. Discuss with students the notion that the class can be considered a population of the small groups. Draw a diagram similar to the one shown below to illustrate this relationship.

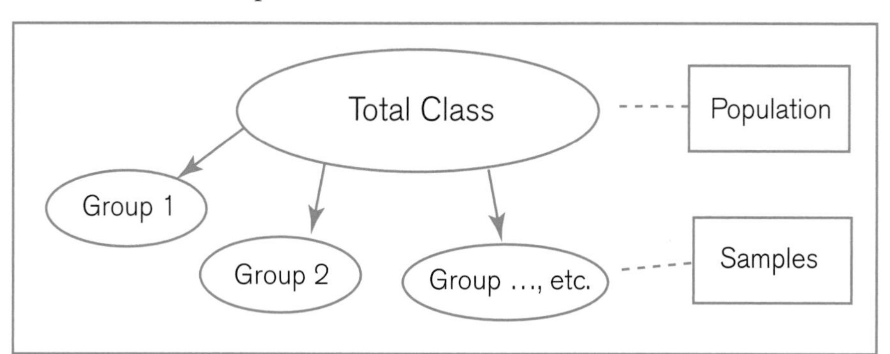

Then record on the board or an overhead transparency the foot length of each student in the class. Ask each student to create a line plot of the class's foot lengths on a copy of the blackline master "The Foot, the Whole Foot, and Nothing but the Foot—Class Data" and answer the questions. Using a transparency of the same blackline master on the overhead projector, record the class data on the line plot and engage the students in a discussion of the questions that they answered on their copies of the blackline master. Record the numerical data that students share in response to the questions on the blackline master.

Have the students compare their responses on the blackline master of class data with their responses on the blackline master of group data by asking the following question: "How close was your sample to the foot length of the total class population?" If the class's median and range are similar to (e.g., within one to two centimeters of) what they observed in the sample, they will probably be satisfied that the sample was representative of the population.

If the class's mean and range were not close to those of the sample, ask the students to reflect on the sample of the class that was selected for the discussion of foot length. Have the students make some hypotheses about what possible factors could have *biased* the sample in such a way that the sample might not have been representative of the class. Students may have noticed that the "counting off" process resulted in a sample consisting of the tallest (or shortest) people in the class or a group of all boys (or all girls).

A biased sample *is a sample in which some portion of the population has a higher proportion of members than actually occurs in the population.*

Extend

Ask the students, "If we survey a random sample of the *entire school* (population) on the length of the students' feet to the nearest centimeter, will the line plot be similar to the line plot that we have made for our *class*? Why, or why not?" Some questions to stimulate the groups' thinking might include these: "Will the shortest foot lengths on the line plots be about the same?" "What about the longest?" "Will the ranges of measurements be the same?" "Will the modes be the same?" and "How will the medians and the means of the distributions compare?"

Explain to the class that one way to select a random sample of the entire school (population) would be to put all the students' names in a hat and select a portion of the names. Ask the class, "In a data investigation, when you choose a sample, how can you be sure that the data are representative of the population?" Have the students work in their groups to answer the question by generating strategies that they would use to identify a random sample of students in their school. Ask each group to share one of its ideas with the class and have the class decide whether the method would generate a random sample or not. Students might suggest that they could obtain a list of all the students in the school and systematically select every sixth, tenth, or other-numbered person on the list. They might also suggest dividing the school population into groups according to some characteristic that is important to the study, such as age or grade, and then randomly selecting participants from each of these smaller groups (thus creating a *stratified* sample).

A random sample *is a sample created by a selection process in which members of a population or have an equal chance of being included in the sample.*

Students may choose to conduct the survey of a random sample of the school and then make inferences about the length of the feet of the school population.

Assessment Ideas

When assessing students' understanding of the concept of sample, begin to notice whether students recognize that different samples from the same population can vary. You want students to begin to discover that predictions based on samples may vary.

Where to Go Next in Instruction?

As a result of this activity, students should become more comfortable making inferences about a population on the basis of their knowledge of a sample. In the next activity, Can You Catch Up? which involves an *experiment*, students will have an opportunity to make inferences and predictions about how far ketchup will run on a plate in a variety of timed ketchup races.

Can You Catch Up?

Grades 3–5

In this activity, students will have opportunities to predict whether different temperatures affect the distance that a substance (ketchup) will flow during a 30-second period. Students will make predictions and record the length of the ketchup flow for each temperature. Students will construct a bar graph to display their predictions compared with the experimental results and will engage in a discussion of the variables that may have affected the differences.

Goals

Students will—

- predict the length of the flow of ketchup that is hot, cold, and at room temperature;
- conduct an experiment to determine the distance that the ketchup flows at different temperatures;
- design further investigations to evaluate their conclusions.

Prior Knowledge

Students should have experiences in collecting, organizing, displaying, and analyzing data. They should have a working knowledge of positional vocabulary, including *horizontal* and *vertical* as associated with bar graphs, and be able to measure lengths to the nearest centimeter.

Materials and Equipment

- Three packets of ketchup for each pair of students (or three plastic bottles of the same brand of ketchup)
- One container filled with ice and water
- One electric coffeepot or other container that will keep water hot (such as a Crock-Pot or a thermos filled with hot water)
- One container of water at room temperature
- One nine-inch plastic plate for each pair of students
- One metric ruler for each pair of students
- One permanent felt-tip marker
- One washable felt-tip marker for each pair of students
- Cardboard wedges for each pair of students, cut at a 45-degree angle with the angle marked
- One copy of the blackline master "Can You Catch Up?" for each pair of students
- A stopwatch or a watch with a second hand
- Paper towels
- Kitchen tongs

Show the students how to position their plates against the wedge (see fig. 3.2).

p. 110

This activity has been adapted from Cothron, Giese, and Rezba (1996, pp. 1–8).

Classroom Environment

In the first phase of this activity, you will demonstrate where to place the different types of ketchup on the plate, how to hold the plate at a 45-degree angle during a ketchup race, and how to construct a bar graph to represent both the predictions and the actual lengths of the ketchup flows for the three different temperatures of ketchup. Students will work in pairs to carry out their own investigations to test their hypotheses. Finally, the students will continue to work in pairs to construct bar graphs of the results.

Activity

To prepare for this experiment, fill a container with tap water and allow adequate time for the water to reach room temperature. Fill a second container with ice water (both ice and water should be in the container). The third container should be a coffeepot, a Crock-Pot, or other device capable of heating water and keeping it warm. Each of your containers should be large enough to accommodate a packet of ketchup for each pair of students or a plastic bottle of ketchup that you will use for the whole class. Place the ketchup packets or bottles in the water baths that you have prepared at the different temperatures. Leave them for at least fifteen minutes to give the ketchup time to change its temperature. When using the hot water near students, caution them appropriately to prevent injury. Be sure to remove the ketchup from the hot water yourself. Use tongs to retrieve packets of heated ketchup. Be extra careful when removing a whole ketchup bottle from hot water, and be sure to have paper towels handy for wrapping the bottle.

Prepare a nine-inch plastic plate for each pair of students. On the plate, draw a starting line (see fig. 3.1) approximately two inches from the edge of the plate, using a permanent felt-tip marker. Above the line, write C, R, and H to stand for *cold*, *room temperature*, and *hot*, respectively. Also, prepare a 45-degree-angle cardboard wedge for each pair of students by taking a six-inch square piece of cardboard and cutting it along the diagonal.

Fig. **3.1.**

Prepare plastic plates and cardboard wedges for each pair of students

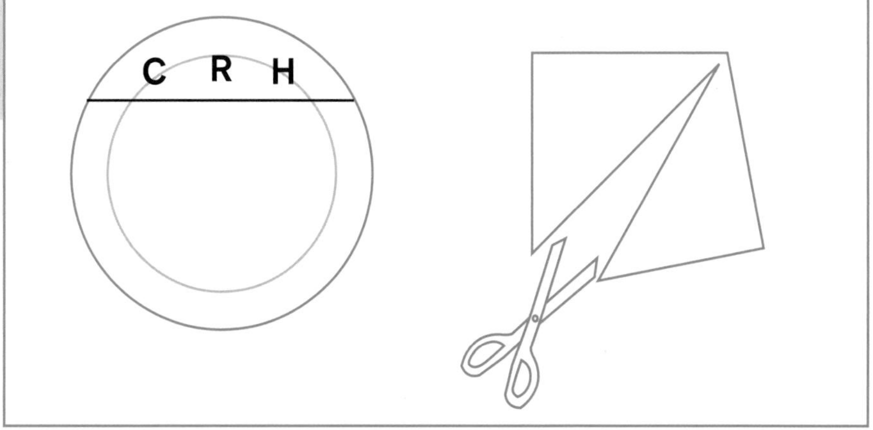

Engage

Ask students, "In a thirty-second race, will the temperature of the ketchup affect the length of its flow down a tilted plastic plate?" Encourage students to think about whether there will be differences in the flow of cold, room-temperature, and hot ketchup. Hand out a copy of the blackline master "Can You Catch Up?" along with one premarked plastic plate, a washable marker, and a precut 45-degree-angle cardboard piece to each pair of students. Tell them that they will hold ketchup races. Ask each pair to use a ruler and their marker to draw three lines perpendicular to the line on the plate to indicate the distance they predict that the ketchup will run in thirty seconds when the plate is held at a 45-degree angle to the tabletop. They should do this for each temperature of ketchup under each letter, C, R, and H. If students are not familiar with angle measure, display a plate held at a 45-degree angle on the tabletop (see fig. 3.2). Have the pairs record the length that they predict for each temperature in the table on their blackline master.

Explore

Test the room-temperature ketchup first. If you are using packets of ketchup, distribute the room-temperature packets to the pairs of students. With their plates lying flat on the table, the students should squeeze their entire packet of ketchup just above the line under the R. If you are using a bottle of ketchup, you should squeeze or pour approximately one tablespoon of ketchup on each plate (a glob the size of a quarter) above the line under the R. Tell the students that at the command of "Ready!" they should tilt their plate against the 45-degree-angle cardboard piece. Say "Go!" almost immediately and let the students watch the ketchup race. Tell the students to lay their plates down on the table as soon as you say "Stop!" after thirty seconds.

Have the students measure and record in the table on their copy of the blackline master the distance, to the nearest centimeter, that the room-temperature ketchup flowed. Repeat this procedure for the cold and the hot ketchup.

Fig. **3.2**

A plate held against a cardboard wedge cut at a 45-degree angle

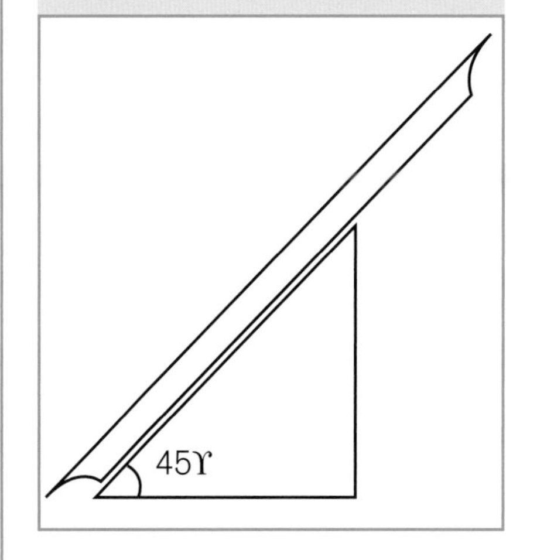

Review with students the construction of a bar graph. Call their attention to the following elements of the graph that they will be making on the blackline master:

- The categories are listed on one axis (most often found on the horizontal axis).
- A numerical scale of equal increments is represented on the other axis. This scale may begin with zero.
- The bars representing each category have the same width.
- There are spaces between sets of bars for ease in readability.
- The height of each bar represents the distance the ketchup has flowed.
- The top of the graph shows a title.

Allow students time to determine the differences between their predictions and the results of their ketchup races. Ask the students, "Did your results support or not support your predictions?" Tell students that if this experiment were only the first, scientists would need to do more experiments to determine the effects of temperature on the length of the path of the ketchup. Ask other questions: "Why were some of your predictions accurate and others not?" "What are some of the factors that may have influenced the results?" For example, some students may have predicted that the hot ketchup would flow faster than the

An analysis of an experiment can lead students to determine that a wide variety of variables may affect the outcome of their investigation. Encourage students to investigate the constants, or the things in the experiment that remain the same or are kept the same, as well as the variables, or the things that change or could be changed.

other two, expecting the heating process to have made it runny in comparison with them. However, some students may have found that the cold ketchup slid down the plate in a glob instead of flowing, suggesting that the cold ketchup traveled a longer distance.

Suggest other variables that might have affected the outcomes: "Do you think that the results of our investigation would have been different if we had used other brands of ketchup?" "What do you think might have happened to the results of our investigation if you had tilted the plastic plate at a different angle?" Ask students to answer the following questions and make a prediction: "If our class were to repeat this investigation, would the outcomes be the same or different? Why?"

Focus the discussion on the variables that may have affected their predictions and on the actual outcome. Encourage students to consider that the consistency of the ketchup, the brand of ketchup, the angle of the plastic plate, and the temperature of the ketchup may influence the outcomes. Ask the students to think of at least one question that other students could answer by analyzing the bar graph, and tell them to write their question at the bottom of their copy of the blackline master "Can You Catch Up?"

Extend

A possible extension to this activity would be to graph the length of the ketchup flow in thirty seconds for different brands of ketchup when experimenting with only *one* temperature type. Analyzing the flow of different brands of room-temperature ketchup in the context of a problem or question may yield different results according to the preferences of the investigator. One such question might be "How do you like your ketchup when using it with french fries or on top of a hamburger?" Students may also recognize that flow is not the only factor to consider when making this selection; taste, color, and aroma are other relevant factors. Students may also wish to use the same procedures to predict and collect data on the flow of mustard. Fourth- and fifth-grade students may wish to represent the results of their investigation on three line plots, with each line plot representing the actual lengths of the ketchup flows found by all the students for the three temperatures of ketchup. Once all the data have been collected for each of the temperatures, have the students determine the median and range of the ketchup flows for each temperature.

Assessment Ideas

When assessing whether students understand the difference between a prediction and the actual results of an experiment, pay close attention to those students who want to go back and change their prediction to the actual result. Although those students may want to proclaim that they made a perfect prediction, they need to understand that a prediction is a guess based on previous experiences and knowledge and does not always come close to the actual result. Those students should begin to recognize that the variables they considered in making the prediction are often the same variables that they need to consider when determining what affected the actual results.

When reviewing their bar graphs, students will begin to see patterns in their original thinking. For example, students will begin to learn what to look for when analyzing their own graphs if you point out such things as the similarities and differences between their predictions and their actual results for the three temperature levels. You might say, for example, "The graph shows that for all three temperature levels, you predicted the flow of ketchup to be greater [or less] than the actual flow."

Where to Go Next in Instruction?

Students should have experiences making inferences and predictions using sampling techniques and experimentation. The next activity, Chores—How Many Hours a Week Are Typical? gives students an opportunity to use these skills in the context of gathering data on the typical number of hours spent on chores by fifth graders in a class in order to make predictions about the general population of fifth graders.

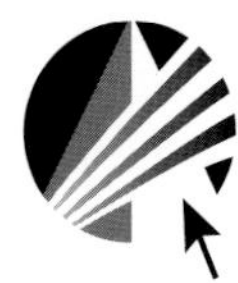

"Propose and justify conclusions and predictions that are based on data and design studies to further investigate the conclusions and predictions." (NCTM 2000, p. 176)

Chores–How Many Hours a Week Are Typical?

Grade 5

This activity builds on the students' experience with the relationship between a sample and a population in the first activity in this chapter, The Foot, the Whole Foot, and Nothing but the Foot! and demonstrates the model of random sampling with replacement as students answer a question concerning the number of hours that they spend each week doing chores. Groups of students will use a survey to collect and organize data on the number of hours a week spent on chores by fifth graders in their school. Then they will use a stem-and-leaf plot to graph the data. Each group will infer what is the typical number of hours a week spent on chores by fifth graders on the basis of data that they collect from a random sample of fifth graders within the school. Students will have an opportunity to compare their results with those of other groups. From their discussions, students will learn (*a*) to recognize the value of using a sample as an estimator for a population and (*b*) to identify potential causes of variation in the results among samples.

Goals

Students will—

- recognize that a sample statistic is an estimation of a population parameter;
- recognize that data collection must be well planned to account for factors causing variations among samples;
- observe measures of central tendency, the spread of the data, and the effect of outliers on their data in the context of a real-world problem;
- describe how errors may present themselves in the course of the data collection process;
- make observations from the data and then develop and evaluate inferences and predictions.

Prior Knowledge

Before this lesson, the students should have had experiences in identifying measures of center (mean, median, and mode), in recognizing trends in line graphs (increasing or decreasing), and in describing the difference between a sample and a population.

Materials and Equipment

- Calculators
- One copy of the blackline master "Chores—How Many Hours a Week Are Typical?" for each student participating in the survey (including those in at least two other classes)
- A copy of the blackline master "Stem-and-Leaf Plot of the Group's Sample Data" for each group of students

pp. 111, 112

- Two transparencies of the blackline master "Sample versus Population—How Do They Compare?"
- A copy of the the blackline master "Sample versus Population—How Do They Compare?" for each group of students

p. 113

Classroom Environment

Students work in groups of five to six on the activity.

Activity

Engage

Introduce this activity with a motivating question: "How much time a week do you think other fifth-grade students spend on chores? Do you think they spend more time or less than you?" Tell the students that you want them to work in groups to collect data from a random sample of students in all the fifth-grade classes in their school (population) to answer the question "How much time does the typical fifth grader spend each week on chores?" Ask the students to predict what they expect will be the average number of hours (rounded to the nearest quarter hour) that fifth-grade students spend on chores in a week. Then have them record their prediction so that they can compare it later with the actual data that they collect. (*Note:* In order to ensure a large enough population for this data collection, the students should collect data from at least two other fifth-grade classes. Where this is not possible, you may wish to change the question to "How much time do typical fourth and fifth graders spend on chores?" and have the students gather data from equal numbers of fourth and fifth graders.)

If all the students who are going to be surveyed can keep daily records for one week before the activity, your students will be able to gather data that relies on more than students' imperfect memories. You can ask your own students to keep such records, and you can enlist the cooperation of the other teachers in doing the same with the other students whom your students will be surveying.

Have the groups use the following plan to collect data on the number of hours that a random sample of students from the population of fifth graders spends weekly on chores:

- Conduct a survey of all students in the fifth grade, including the students in your own class. Ask each student to complete the survey form (Student Survey) on the blackline master "Chores—How Many Hours a Week Are Typical?" or an alternative survey form designed by the students to collect four items of data from those surveyed:
  1. Their age
  2. Their gender
  3. Their class (e.g., Mrs. Jones's class)
  4. The number of minutes a week that they spend on chores

The sample survey on the blackline master is designed to identify the areas of work in the home that are considered to represent chores. By using predetermined chores, you are more likely to ensure some consistency of responses. The students who are conducting the survey should convert the number of minutes supplied by each person completing the survey to the number of hours. To do this, students should divide by 60 (60 minutes in an hour) and express the time in decimal format, with the quotient rounded to the nearest tenth. (e.g., 370 minutes divided by 60 minutes in an hour = 6.2 hours).

- Place each of the surveys in a box (population).

- Have each student group in turn select fifteen surveys from the box. These surveys will make up a random sample of students from across all the classes surveyed. The group should record the data from their fifteen surveys and then replace the survey forms in the box (and mix the contents of the box) before the next group draws its random sample.

This method, known as *sampling with replacement*, guarantees that the probability of a person, object, or event being selected by each group remains the same throughout the sampling process. Thus, each survey form has an equal chance of being selected whenever a random sample is drawn.

Ask each group to create a stem-and-leaf plot on the blackline master "Stem-and-Leaf Plot of the Group's Sample Data" to display the group's data. Have the students compute the mean, median, and mode for their sample. Since stem-and-leaf plots are not limited to two-digit data, students should be shown how to use the stem to represent the number of hours in the quotient and the leaves to represent the decimal places in the quotient (i.e., .0, .1, .2, .5, etc.). Students should separate numbers in the leaves by commas to avoid misinterpretation of the numbers.

Explore

Using their stem-and-leaf plots representing their random samples of the typical number of hours of chores of a fifth grader, the groups should report their findings to the class by discussing their answers to the following questions: "What did you find?" "What is the mode, or the most common number, of the hours of chores in the sample you surveyed?" "What is the median number of hours of chores in the sample you surveyed?" "Predict the average number of hours of chores, and then, using your calculator, determine the average (mean) number of hours of chores in the sample you surveyed. Was the average what you predicted?" "What is the range of the number of hours of chores in the sample you surveyed?" "What does this tell you about the distribution of the values?"

A data set containing numbers of hours may be presented in a stem-and-leaf plot as shown in the following example: Sample Data Set: 2.5; 3.6; 5.0; 3.1; 2.6; 4.3; 5.2; 3.3.

Hours	*Part of Hour*
2	.5, .6
3	.1, .3, .6
4	.3
5	.0, .2

Have the students critique the survey results of the sampling process. After reviewing the terms *population*, *sample*, and *random sample* in relation to the data, ask the students to judge the effectiveness of the sampling process: "Does this sample represent an estimate of the population of fifth graders?" "Why, or why not?" Encourage students to think about the issues that affect the "representativeness" of their samples and how well each sample represents the larger population by asking, "Does the random sample represent the population?" Ask the students to explain why or why not. Engage the students in thinking about other methods to collect the data randomly, by asking: "What other ways could your group have collected data on a random sample?"

Ask the students whether a graph other than the stem-and-leaf plot would have been easier to use to interpret the data on the number of hours of chores: "Was the stem-and-leaf plot easy to use to understand the information that was presented?" "Would another type of graph have been more appropriate for presenting these data?" "What could you have done differently in your data collection that would have changed the graph you used to present your data?"

Using a transparency of the blackline master "Sample versus Population—How Do They Compare?" write the number of hours of all students who were surveyed (the population) on the stem-and-leaf plot. (*Note:* If the total number of fifth-grade students in your school makes this task a very time-consuming assignment, think about the following alternative: Tell your students to consider their class as a population, with the boys making up one sample and the girls another sample. The students can then compare each group—that is, boys and girls—with the population, or total class. In addition, the students can perform cross-gender comparisons to determine differences and similarities. Be sure to note, however, that "all the boys" and "all the girls" are not random samples.) You may wish to have students take turns removing the surveys from the box, one at a time, and reporting the information. To avoid confusion, students should discard the surveys after they are recorded. Rewrite the stem-and-leaf plot on the second transparency so that the data are presented in a sequential, organized manner.

Ask the students to work individually to answer a similar set of questions about the population, including the following: "What is the median number of hours of chores for the fifth graders?" "What is the average (mean) number of hours of chores for fifth graders?" "What is the mode, or the most common number, of hours of chores worked by fifth graders?" "What is the range of the number of hours of chores performed by fifth graders in this population?"

Call on students to identify the mean, median, mode, and range for the population and have students check their own work to ensure that they are using the correct computations and processes to identify these measures.

Extend

Ask the students to return to their groups, and have each group focus on how well their randomly selected sample mirrored the population. Ask each group to set up a back-to-back stem-and-leaf plot to compare the data on a copy of the blackline master "Sample versus Population—How Do They Compare?" In a back-to-back stem-and-leaf plot, the stem is in the middle. The leaves are on both sides of the stem. The data for the population should form leaves on the right-hand side of the back-to-back plot; the data from their sample should form leaves on the left-hand side of the back-to-back plot. Students should read from the middle to the right for the data on the population, and they should read from the middle to the left for the sample.

Ask the groups to compare their sample statistics with the statistics for the population. They should realize that although there can be variation between a sample and a population, the larger the size of the random sample, the more accurate it will be to predict information about a population. Students should recognize that random sampling and the methods of sampling that they use can significantly affect their predictions about a population.

Students should be encouraged to design further studies to investigate their conclusions or to examine other conjectures or predictions. Are there other questions or topics that they can investigate? Are there other questions that they can pose? For example, the students may wish to pose this question: "Is there a difference between third and fifth graders in the typical number of hours spent on chores?" Other topics

Two data sets containing numbers of hours may be presented in a back-to-back stem-and-leaf plot as shown in the following example:

Data sets (number of hours):

Boys	Girls
2.1	5.3
3.3	2.2
2.5	3.6
1.9	4.1
3.7	3.2
2.2	2.9
4.6	3.1
2.7	3.3

Back-to-back stem-and-leaf plot:

Boys Parts of an Hour	Hours	Girls Parts of an Hour
.9	1	
.7, .5, .2, .1	2	.2, .9
.7, .3	3	.1, .2, .3, .6
.6	4	.1
	5	.3

that you may wish to use with your students to reinforce the concepts and skills presented in this activity may include these: typical bedtimes, arm spans, times that it takes students to run the fifty-yard dash, numbers of curls or pull-ups that they can do in one minute, and numbers of minutes that it takes them to walk or run a mile.

Assessment Ideas

Create a prediction bulletin board and give bonus points to students who post information that involves a prediction about a situation. The information can be about something that they have observed themselves or an item such as an article or news clipping in which predictions play a part.

Where to Go Next in Instruction?

The question "How representative is a sample statistic of a population parameter?" has probability as its conceptual basis. This chapter focused on ideas and instructional strategies to give students opportunities to explore notions of statistical inference and prediction. The next step is to investigate ideas of probability. The following chapter presents activities to introduce students to the basic notions of chance and "likelihood, " as well as initial explorations in probability processes.

GRADES 3–5

DATA ANALYSIS *and* PROBABILITY

Chapter 4
What Are the Chances?

This chapter focuses on probability ideas, or the mathematics of chance. We use probabilistic reasoning every day when we consider the chance of rain or the likelihood of our favorite team winning a baseball game. Probability is the study of the likelihood, or chance, that a specific event will happen. The study of probability will help students to develop problem-solving and reasoning skills.

The activities in this chapter will help students develop initial ideas in probability, such as that the probability of an event occurring can be expressed as a ratio of the number of times that the event can occur to the number of all possible outcomes in a given situation. The activities emphasize experimentation and use relative frequencies to determine the likelihood of an event. In How Likely Is It to Land in the Trash Can? students first associate the likelihood of a paper toss landing in the wastebasket (the outcome of interest) with terms such as *likely* and *unlikely*, and then they connect the probability of each type of toss with a value from 0 to 1. In the second and third activities, Is There Such a Thing as a Lucky Coin? and Spin City, students predict the probability of outcomes in simple experiments—one using coins and the other using spinners—and then test their predictions by gathering data and finding the experimental probabilities of the events. They examine class data, make line plots, and discuss the variability of outcomes. In the final activity, Is It Fair? students explore a game in which the likelihood of winning is not the same for each player. This leads students to think mathematically about the meaning of fairness and to change the game to make it fair.

How Likely Is It to Land in the Trash Can?

Grades 3–4

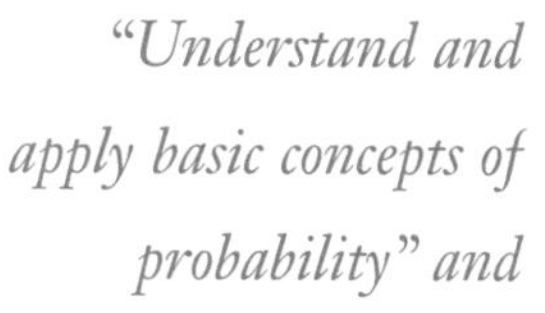
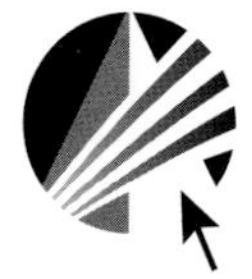

"Understand and apply basic concepts of probability" and "describe events as likely or unlikely and discuss the degree of likelihood using such words as certain, equally likely, *and* impossible." *(NCTM 2000, p. 176)*

This activity addresses the *likelihood* of an event happening. Students toss a crumpled ball of paper into a trash can from varying distances after predicting whether their toss is certain to go into the can (1), likely to go in, equally likely to go in and not to go in (1/2), unlikely to go in, or has no possibility of going into the can (0). They conduct many trials, collect and organize the data, and then as a group classify the likelihood of the paper landing in the can from each distance. They quantify the likelihood by assigning values from 0 to 1. They then generate lists of other events that are certain, likely, equally likely and unlikely, unlikely, or impossible and place them on a probability number line. As an extension they conduct another experiment and assign values of 0 through 1 to represent the likelihood of the different outcomes.

Goals

Students will—

- describe an event as certain, likely, equally likely to occur and not occur, unlikely, or impossible;
- quantify the likelihood of an event using a value from 0 to 1.

Prior Knowledge

- Students should be able to represent a part of a whole or a part of a set using fractions.
- Students should be able to place 1/4, 1/2, and 3/4 correctly on a number line.
- Students should have had experience using a metric ruler.

Materials and Equipment

p. 114

- Sheets of scrap paper or recycled paper—all the same size
- Trash cans or other large containers, such as buckets
- Metric tape measures or metersticks
- Masking tape
- Pad of sticky notes
- A copy of the blackline master "Paper Toss Recording Sheet" for each group of three students

Classroom Environment

Students work in groups of three to gather data. A whole-class discussion follows the data collection.

The idea for this lesson is adapted from Singer, Konold, and Rubin (1996).

Activity

Engage

Ask students if they have ever crumpled a piece of paper and aimed it at a trash can. Did their tosses go in? Ask, "Can we predict whether or not a toss will go in the trash can?" Allow a brief discussion of this question, since many children will state that they can always "make" a toss. Next ask, "Why don't tosses go in 100 percent of the time?" Possible reasons might be that the student was a long distance from the trash can, that there were obstacles in the way, that the student was not very good at throwing, or that the paper wasn't crumpled tight enough and didn't fly smoothly through the air. Introduce the idea that some tosses will be "certain" to go in, whereas others may be just "likely" to go in. Continue by stating that other tosses may or may not go in and that there are tosses that are unlikely to land in the trash can or that cannot possibly land in it.

An event that is certain is one for which the probability of the event occurring is 100 percent, or 1, and an event that is impossible is one for which the probability of the event occurring is 0 percent, or 0. The term *50-50* is sometimes used to indicate that an event has a 50 percent probability of occurring and a 50 percent probability of not occurring. Saying that the occurrence and nonoccurrence of an event are *equally likely* also indicates that the likelihood of the event occurring is half (1/2, or .5, or 50%). The descriptors *likely* and *unlikely* do not have specific numerical values associated with them. *Likely* describes an event that has a probability between 1 and .5 (or between 100% and 50%) and *unlikely* describes an event that has a probability between .5 and 0 (or between 50% and 0%). Students may use the descriptor *maybe* (which is an informal term) to describe probabilities that are around 50 percent.

Write *certain, likely, equally likely to occur and not occur* (or *50-50*), *unlikely*, and *impossible* on the board and ask different students what each term means to them. Be sure to have two or three students talk about each of the terms so that you can assess students' initial understanding of them. For now, you are simply trying to find out what your students think about these terms. Explain that the students will be experimenting to determine the likelihood of a paper toss landing in a container and that as a class you will all come to some agreement about the meanings of the terms.

Explore

Divide the students into groups of three and give each group a trash can or another large container (such as a bucket), a measuring tape, and a copy of the blackline master "Paper Toss Recording Sheet." The students will be making tosses from the following distances to the trash can: 3 cm, 150 cm, 250 cm, 350 cm, and 6 m. Tell students that when tossing from the 6-meter distance, they will close their eyes! Before tossing crumpled paper into the trash can, students must measure the different distances from the trash can and then mark them by placing a small piece of masking tape on the floor. You will want to check that students are using the measuring tools correctly and that they understand how to measure the distances, such as 350 centimeters. Once students have marked the distances, they will use their copy of the

The trash cans or other containers should be the same size for all groups. However, more than one group of students can use one receptacle.

If your students have worked with decimal numbers, you can introduce the distances in meters–0.03 m, 1.5 m, 2.5 m, 3.5 m, and 6 m.

You may wish to assign group jobs to make the collection of data run smoothly. For example, while one student is making his or her tosses, another can be retrieving the crumpled paper, and a third can be recording the data on the recording sheet.

blackline master "Paper Toss Recording Sheet" to record what they predict as the likelihood that their tosses from each distance will land in the container. They will then each toss a crumpled piece of paper three times from each distance and record the number of tosses that land in the basket. The longest toss, 6 meters, is further complicated because students are to close their eyes when making those tosses.

Before actually making the tosses, have the class discuss the likelihood of the tosses going in from each of the distances. Write the distances on the board; next to each distance, write the descriptor (e.g., *certain, unlikely*) that the class as a whole agrees best describes the likelihood of the toss landing in the container. There may not be agreement at this point, so some distances may have several different descriptors next to them.

Once students have completed tossing crumpled paper, combine the class data on a chart, as in the sample below.

Distance	Number of Tosses That Went In	Total Number of Tosses	Tosses In/ Total Tosses	Descriptor
3 cm	60	60	60/60	
150 cm	51	60	51/60	
250 cm	35	60	35/60	
350 cm	12	60	12/60	
6 m (eyes closed)	0	60	0/60	

Next, initiate a discussion about the data. The goal is to help students agree on descriptors for each likelihood and to notice the relationship between the attempts and the successes from each distance. If students' initial predictions about the likelihood of different tosses were quite different from the actual data, the follow-up discussion will enable you to help them resolve the conflict between what they predicted and what occurred. If 51 out of 60 tosses from 150 centimeters went into the can, you might ask students to describe the relationship between the number of tosses that went in and the total number of tosses. Students might suggest that "a lot more than half of the tosses went in," or use a fraction in expressing the relationship (for example, "more than 3/4 went in," or "about 5/6 of the tosses went in," depending on prior instruction.

Another goal is to help students associate the benchmark of one-half with "equally likely and unlikely," "equally likely to occur and not occur," or 50-50. Ask students questions such as "About how many tosses went in from 250 centimeters?" and "Why do you think we say the likelihood is 50-50 when about half of the tosses landed in the can?" Students might respond that 50-50 suggests that about half the time a toss will go in and about half the time the toss will miss. A likely chance implies that more than half but not all the tosses will go in—and the closer the number of successful tosses is to the total number of tosses, the more likely the event.

You might hear students using the informal term *maybe* to represent the probability of an event that is equally likely to occur and not occur. Encourage them to think of such an event as one whose occurrence and nonoccurrence are *equally likely* or as having a likelihood of *50-50.*

Although the data in the chart show that all the attempted tosses went in from 3 centimeters, you might have a situation where someone missed from that distance. This is a wonderful opportunity to clarify the concept of "certain." When an event is certain, it will definitely happen—for example, the sun will always rise in the east. You may wish to have students consider how to adjust the distance so that a toss will be certain to go in—perhaps by holding the paper directly over the can! The same considerations must also be given to events that are impossible—these events can never happen. Since it is not impossible (just unlikely) for someone's toss from 6 meters with his or her eyes closed to land in the trash can, you might want the students to discuss parameters that make tosses impossible. For example, tossing a crumpled piece of paper in a trash can from a distance of 1 kilometer with eyes closed might be suggested as an impossible event.

Next, draw a long line segment on the chalkboard and label it as shown below.

0	1
Impossible	Certain

Explain that this is a probability line and that we can use it to quantify the likelihood of events. Probability is a number that indicates how likely something is to occur. Events that are impossible are said to have a likelihood of zero (0), and events that are certain are said to have a likelihood of one (1). Ask students where on the line we could place events that may or may not occur or that have a 50-50 likelihood? (In the middle, at 1/2.) Suggest that students discuss with the person next to them where they might place likely and unlikely events and what fractional value they might give to each likelihood. After a few minutes, arrange for different students to come up and place a sticky note with the term *likely* or *unlikely* on the line and to explain their reasoning for the placement. There is not an exact spot for these terms on the probability line, and students will place their sticky notes in different spots. However, *unlikely* should be placed between *impossible*, or 0, and 1/2—the closer the sticky note is to *impossible*, the less likely an event with that likelihood would be to occur. Although fractions such as 1/4, 1/5, or 1/6 are often equated with an unlikely event, there is not one fractional quantity associated with the terms *likely* and *unlikely*. We must apply judgment to each situation when assigning likelihood descriptors.

A likely event should be placed between 1/2 and *certain*, or 1; often it is placed approximately at the 3/4 position. Follow up this activity by having students place the paper-toss distances on the probability line (they can write these on additional sticky notes). Your final line might look like the one below. Notice how the class changed the distances for certain and impossible events.

Some students may have predicted that landing a toss from 6 meters with eyes closed would not happen, because it was impossible. This is not true, of course. A discussion of the difference between the meaning of the terms impossible *and* unlikely *and asking students to generate examples of events that are impossible or unlikely might help resolve any confusion.*

Students whose first language is not English may have difficulty with the probability terms mentioned in this lesson. The terms *likely* and *unlikely* can be confused with the verb *to like* (as in "I like ice cream"). Linking these terms to the ratio comparing the number of tosses that land in the trash can to the total number of tosses can help English language learners make connections.

0	1/4	1/2	3/4	1
Impossible	Unlikely	Equally likely to occur and not occur	Likely	Certain
2 km	350 cm	250 cm	150 cm	1 cm above can

Extend

As an extension to this lesson, students can conduct simple experiments in which they draw two counters at the same time from a bag of counters and then put them back in the bag. For the first experiment, they will put two blue and two white counters into a bag. Students will first estimate the likelihood that the two counters that they draw will

If students have studied decimals or percents, you may wish to quantify these terms by using one or both of these representations.

be the same color and then the likelihood that the two counters will be different colors. They should use both word descriptors and fractions to describe their predictions. They should then make at least thirty draws and record their results in a table, chart, or line plot. On a probability line, students should record the result of their draws, using words and fractions. For example, students might find that eleven out of thirty draws were the same color. They might write comments like "only in 11 draws out of 30, or 11/30 of the time, were the counters the same color; so it is unlikely that you will draw two counters that are the same color from the bag. So we should write 'matching colors' at 'unlikely' on the probability line." Next, students will use their experimental data to predict the likelihood of future draws being matching colors. Then students should change the composition of the counters in the bag, now putting three blue counters and one white counter into the bag. They should then conduct the same experiment and reflect on how the new composition affects the likelihood of drawing two identically colored counters.

Experiment 1: Put two blue and two white counters in a bag. Draw two counters at the same time from the bag without looking and record the results. Then replace the counters in the bag and repeat the experiment. Record the results of thirty draws.

(The probability that the two counters will be different colors is 2/3, and the probability that they will match is 1/3. If we label the colored counters and list all possible matches, we get B_1B_2, B_1W_1, B_1W_2, B_2W_1, B_2W_2, and W_1W_2. Four out of six, or two-thirds, of the draws do not match in color, and two out of six, or one-third, of the draws do match.)

Experiment 2: Put three blue counters and one white counter in a bag. Draw two counters at the same time from the bag without looking and record the results. Replace the counters in the bag and repeat the experiment. Record the results of thirty draws.

(The probability that the two counters will be different colors and the probability that the counters will match are the same, or 1/2. If we label the colored counters and list all possible matches, we get WB_1, WB_2, WB_3, B_1B_2, B_1B_3, and B_2B_3. Three out of six, or one-half, of the draws do not match in color, and three out of six, or one-half, of the draws do match.)

You may want to encourage students to design other experiments with counters and use these experiments to estimate the likelihood of events.

Assessment Ideas

One way to assess how well students have grasped the concepts exhibited on the probability line is to ask pairs of students to make their own lines and place different events on them for each descriptor. Ask students also to include numeric values, as shown:

0	1/5	1/2	4/5	1
Impossible	Unlikely	Equally likely to occur and not occur	Likely	Certain
We will see two suns in the sky.	It will snow in May.	We get a longer recess.	We will have homework.	The sun will rise in the east.

You can also assess students' understanding of the likelihood of events when you have them discuss their probability lines as a class. You may find that students have more difficulty coming up with examples for one descriptor than another. If so, encourage students to share their examples and to explain their reasoning for choosing those examples.

Where to Go Next in Instruction?

Look for situations in the everyday routine of the classroom where you can ask about the likelihood of an event occurring. You may wish to keep a probability line posted on which students can add events under each descriptor.

Once students understand the meaning of probability descriptors such as *likely* and *impossible* and are able to place the likelihood of events on a probability line, they will benefit from conducting simple experiments in which they determine likelihoods. The next activity, Is There Such a Thing as a Lucky Coin? presents coin-tossing experiments that require students to revisit terms. Students must differentiate between terms of likelihood and more common words such as *lucky* when describing the outcomes of multiple tosses.

Is There Such a Thing as a Lucky Coin?

Grades 3–4

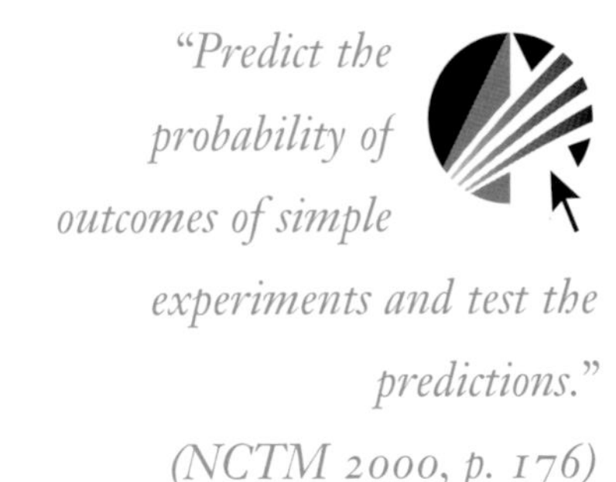

"Predict the probability of outcomes of simple experiments and test the predictions."
(NCTM 2000, p. 176)

In this activity, students learn how to conduct a simple experiment involving coin tossing. They collect and organize their coin data, and use those data to determine simple experimental probabilities. In particular, students discuss what it means in literature for a coin to be "lucky" (usually such a coin lands consistently in a particular way—say, heads up, thus bringing good fortune to the owner, who is able to predict the outcomes). The students then consider whether or not lucky coins really exist. In general, "lucky" coins are not fair coins; they are weighted in such a way as to land either always heads or always tails. In the following activity, students toss pennies, nickels, dimes, and quarters and collect classroom data on the number of heads and the number of tails for each type of coin. They discuss whether or not one type of coin is "luckier" in the number of heads that tossing it produces than another type of coin. Finally, students record the experimental probabilities as fractions and compare outcomes for each type of coin.

Goals

Students will—

- explain why the term *lucky* is not a descriptor of the likelihood of an event, by thinking about the fact that a lucky coin is not a fair coin;
- realize that with a fair coin heads and tails are equally likely to occur and that characteristics such as the size and weight of different fair coins do not affect toss outcomes;
- determine that the experimental probability for getting heads or getting tails when tossing different coins is approximately one-half;
- predict how often an event will happen in a given number of trials.

Prior Knowledge

- Students should be familiar with the idea that fractional quantities can be used to quantify the likelihood of an event.
- Students can describe the likelihood of events by using terms such as *impossible*, *unlikely*, *maybe*, *likely*, and *certain*.

Materials and Equipment

- One penny, nickel, dime, and quarter for each pair of students (the coins must be real!)
- Carpet squares or foam mats for deadening the sound of coin tosses (optional)

Classroom Environment

Students work in pairs for the first part of the lesson. A whole-class discussion occurs in the second half of the lesson.

Activity

Engage

Start the activity by posing a question to students: "Have you ever heard of a 'lucky' coin?" Many students will have read stories or books in which a lucky coin or object helped an individual get through a difficult time or overcome an obstacle. In most of these situations, the coin is believed to bestow luck on an individual who holds it or carries it in a pocket but does not necessarily toss it. Other students may have heard the word *lucky* used for a coin that always lands heads (or tails). Bring up the fact that there are coins that are called "fair" and ones that are "not fair "—namely, coins that are weighted in such as way that they regularly land heads up. Perhaps it is this predictability that makes people call them lucky! Some students may associate "lucky" with "likely." Ask students to explain why they think the two terms are synonymous. Be sure to clarify the difference between the terms if students are still confused.

The dictionary defines the word *lucky* to mean favored by luck, or fortunate. Although the word *likely* has numerous meanings, the mathematical definition indicates that it is a measure of probability between one-half and one.

Explain that the students are going to toss some coins and investigate whether they come up heads or tails more often. Follow this announcement with questions: "Do you think that one type of coin might land heads up more often than another?" and "Are all U.S. coins fair coins?" Possible responses include the following: "Since some coins weigh more than other coins, they will land heads up more often," or "Maybe since some coins are a lot smaller, like dimes, they might not land the same way as a bigger coin, like a quarter," or "No, coins always come up heads half the time and tails half the time."

U.S. coins are fair coins. Although with any fair coin there is an equal probability of its landing heads or tails, this doesn't mean that the sequence of heads and tails from tossing one fair coin will be H, T, H, T, H, T, ... In fact, there is a great deal of variability in the order of the outcomes. Unfair coins are those that are weighted in such a way that they regularly land one way. Direct students to toss each coin ten times and to keep a record of the number of heads and tails for each coin.

Explore

While the students are tossing coins, place a chart that lists the four coins and the number of heads and tails for each (see example below) on the board or overhead projector. As student pairs finish, have them enter their results for each coin on the chart. That is, each pair should record the number of heads and the number of tails obtained after tossing the penny, nickel, dime, and quarter each ten times. The chart below shows possible student responses.

	Penny		Nickel		Dime		Quarter	
	H	T	H	T	H	T	H	T
Student Pair 1	7	3	5	5	6	4	8	2
Student Pair 2	5	5	2	8	7	3	4	6
Student Pair 3	1	9	7	3	4	6	5	5
Student Pair 4	5	5	6	4	3	7	6	4

After the students have recorded their data on your classroom chart, take a minute and have the class members examine their results. Most likely there will be a great deal of variability in the number of heads and tails tossed (as there is in the example above). Ask students to describe

The mathematical notation to show the probability of an event is P(event) = ? The experimental probability of tossing heads in the example here is P(heads) = 74/150.

the total number of heads and tails for each coin. You might ask, "Is there anything surprising in the results?" Possible responses might include the following: "Sometimes the numbers of heads and tails are the same, but not always" and "Everyone didn't toss the same numbers of heads and tails." You may want to post students' questions or observations to return to later. It is important first to total all the tosses and look at the class data as a whole before going back and looking at individual tosses. If you repeatedly toss a fair coin a large number of times, the occurrence of heads will get closer and closer to one-half. When you have a small number of tosses, however, the results may be surprisingly varied. That is, all student pairs will not toss 5 heads and 5 tails—some pairs may have 1 head and 9 tails or 8 heads and 2 tails, and so on. One reason that we add the total number of heads and compare this number to the total number of tosses for the whole class is so that we will have a large number of tosses, or trials. In this example, the numbers will be close to 50 percent heads and 50 percent tails for each coin.

Total the entries in the columns under each coin, starting with the penny. For example, your class totals may be 74 heads and 76 tails out of 150 tosses of a penny. Let's use these data to find probabilities. A probability tells us the relative frequency with which we expect an event to occur. Larger samples give better estimates, and this is why we pool results. We can determine the probability of heads from this experiment by comparing the number of outcomes for the specific result (heads) to the total number of possible results (all tosses). In the penny toss, the probability of heads is 74/150, and the probability of tails is 76/150. Pose questions to students about what you have explained and about these data, such as these: "What is a probability?" "What benchmark fraction is 74/150 close to? How about 76/150?" (1/2) "What words might you use to describe the likelihood of tossing a coin and getting heads?" (*50-50*, or *maybe*, or *about half the time*) "Do you think the two results, heads and tails, are equally likely to occur when you toss a penny many times?" ("Yes. When we look at the class totals, it seems that heads and tails occur about half the time.")

Find totals for the numbers of heads and tails for each of the coins and compare the results. Students will notice that the likelihood of heads and tails is close to, if not exactly, 50-50 for all the coins. Pose the following questions for discussion: "How likely is it that we will get a head when tossing a penny, a nickel, a dime, or a quarter?" "Were you surprised that the results for each coin were so similar?" "Why do you think this occurred?" "Does the size and weight of a fair coin have anything to do with whether it lands heads or tails?" Encourage students to explain their reasoning. You may also want students to explore what happens when the number of tosses is very large—say, 10,000 tosses. Since it would take a long time to make that many tosses by hand, you might want to use a computer program to simulate the results. The outcomes can then be analyzed and discussed. Another question you'll want to pose to bring the discussion back to the original question is "Do any of these coins seem 'luckier' than others, or do they all have the same likelihood of landing heads up?" A possible response might be "Sometimes a coin seems to be 'lucky' because it lands heads up a lot. But then the next time you toss it, it doesn't land heads up. You have to look at a lot of tosses to notice this. I don't think any of the coins are lucky if that means being unfair—here when you add up the number of heads and the number of tails for lots of tosses, the numbers are very close. They all have a 50-50 chance of landing heads or tails."

Next, use these data to make predictions. If students tossed a dime 200 times, about how many heads would they expect? (100) What if they tossed a nickel 1200 times—how many tails would they predict would occur? (600) After discussing many examples using a large number of tosses, ask what they might expect to happen with 10 tosses—"If we toss a coin 10 times, will we get 5 heads and 5 tails?" Suggest that students look at the data in the chart. There will be much disagreement on this issue. Post what students say about the tosses and the distribution of heads and tails on the board as ideas to consider. At this point you are mainly gathering information about students' understanding of distributions. Mention that you will return to this idea at a later date.

Extend

One way to extend this lesson is to use the applet Coin Toss on the accompanying CD-ROM. Students can select the number of simulations that they want of a coin being tossed, and they can decide how they want the results displayed (as a ratio, in a table, or as a list of heads and tails). Students can predict the number of heads in 100, 1000, or 5000 tosses and then compare their predictions to the experimental data generated by the computer. They also can have the applet simulate tossing a coin repeatedly a small number of times (e.g., 10) and use those data to discuss the variability of results.

Assessment Ideas

To assess students' understanding of the lesson, have them answer the following questions:

- "Is there such a thing as a lucky coin that always lands heads? Explain."
- "Josh tossed a fair coin 6 times and got 1 head and 5 tails. Is this possible? Shouldn't he have gotten 3 heads and 3 tails?"
- "If you were to toss a quarter 250 times, about how often would you expect it to land heads? Explain."

You can ask students to answer the questions either as classwork or as homework. If you assign them as classwork, you can have the students work individually or in pairs. If they work in pairs, you can first have them discuss the questions with their partners and then you can follow those pair discussions with a whole-class summary.

You can gain a great deal of information about students' understanding by analyzing their responses to the questions. Their responses may reveal misconceptions and areas that need further discussion or instruction. For example, if students believe that fair coins can be "lucky," consistently landing in a particular way, or if they state that three heads and three tails are the only reasonable outcomes for six coin tosses, you will want to spend more instructional time on the variability of outcomes. You can ask students to conduct additional coin-tossing experiments and to collect more data. You can ask questions like the following: "When we toss a coin twelve times, why aren't the outcomes always six heads and six tails?" However, some responses will indicate that students are making sense of the content. ("Fair coins can't control how they land. Sometimes they land heads up and sometimes they land tails up. The term 'lucky' isn't useful in probability, since the coins can't control how they land." "About 125 of the 250 coin tosses will be heads, since about half of the time heads will occur, especially when I toss the coin a lot of times!")

For additional activities that use experiments to help students consider the likelihood of events, see Edwards and Hensien (2000) on the CD-ROM.

Where to Go Next in Instruction?

Have students investigate the likelihood of other equally likely events, such as rolling a one, two, three, four, five, or six on a number cube. It is important that students not conduct experiments using only coins. Students need many experiences in predicting the outcome of simple experiments and in testing their predictions. They need to work with a variety of models in order to generalize many probability concepts, such as that the likelihood of an event is always represented by a number from 0 to 1. The next activity, Spin City, has students again making predictions and testing their predictions by gathering data. In this task, however, they work with spinners—another common probability model.

Spin City

Grades 4–5

In this activity, students learn that experimental probabilities differ according to the characteristics of the model; they also grapple with the idea of variability—that two identical spinners may not result in identical experimental data. Given a variety of spinners, students conduct experiments and graph the outcomes for each spinner on line plots. Students discuss why different spinners produce different line plots and why different groups of students using the same spinner do not have identical outcomes. They combine group data for the spinners and calculate the experimental probabilities. Students then match spinners with possible resulting line plots of experimental data.

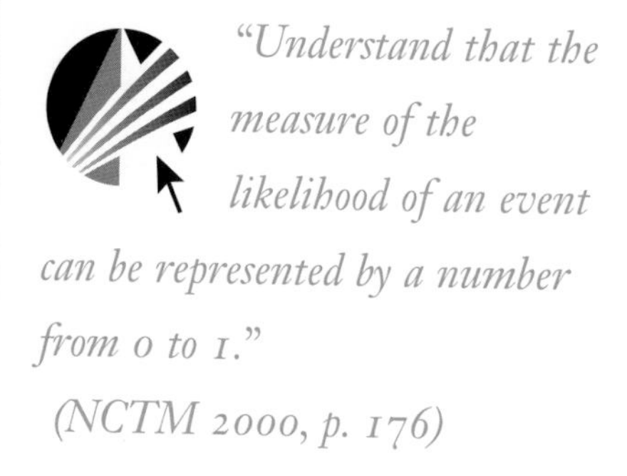

"Understand that the measure of the likelihood of an event can be represented by a number from 0 to 1."
(NCTM 2000, p. 176)

Goals

Students will—

- use line plots to represent outcomes from probability experiments;
- examine the distribution of results in probability experiments;
- investigate situations where the outcomes are not equally likely;
- calculate experimental probabilities for events;
- use probability to predict how often an event will happen in a given number of trials.

This activity uses experimental data collected by students from different spinners to represent the probability of each spinner landing on a specific section as a number from 0 to 1.

Prior Knowledge

- Students should understand that fractions from 0 to 1 can quantify a probability.
- Students should have experience using tallies to record outcomes of experiments, and they should understand that probabilities can be calculated by using data from experiments.

Materials and Equipment

- A clear plastic spinner for each pair of students
- A copy of the blackline master "Spin It" for each pair of students
- A copy of the blackline master "Matching Line Plots with Spinners" for each student

pp. 115, 116

Classroom Environment

Students work in pairs for the entire lesson. They individually match spinners and line plots.

Activity

Engage

Hand out a clear plastic spinner and a copy of the blackline master "Spin It" to each pair of students. Ask them to experiment with aligning the center of the spinners with the center of the circles on the blackline

Clear plastic spinners are readily available from distributors of mathematics manipulatives.

The idea for this lesson is adapted from Singer, Konold, and Rubin (1996).

masters. When everyone can do this satisfactorily, ask them if they have used spinners before, and if so, where? Spinners are most commonly used in games. However, many students will never have used one before. Ask them to examine the circles on the blackline master "Spin It." What do they notice? Both circles are the same size and are divided into five sections, but the sections on spinner 1 are equal sizes and the sections on spinner 2 are different sizes. The sections are labeled A, B, C, D, and E. Explain that the students will be investigating the likelihood of a spinner landing on any one of these sections (A through E) by collecting data and calculating probabilities. Post these directions on the chalkboard and encourage students to begin!

1. For each circle, predict if the spinner is more likely to land on some sections than on others.
2. Out of forty spins for each spinner, estimate the number of spins that will land on section A in each spinner.
3. Spin each spinner forty times, and record the outcomes for each circle by using tally marks.
4. Enter the data for the letters on each spinner into the Class Data Chart on the board.

You will need to put two charts on the board, one for spinner 1 and one for spinner 2, for student pairs to use to record the number of times the spinner landed on sections A through E.

Explore

After the pairs of students have conducted their experiments and recorded their data on the board, call the class together to discuss the results. Since spinner 1 and spinner 2 are different, discuss them separately. Later you can compare and contrast the two experiments. Students' experimental data should point to the fact that on spinner 1, outcomes A through E are equally likely. That is, all the sections are the same size, and each letter will be selected approximately 20 percent, or one-fifth, of the time. Use the data on the board to calculate the experimental probability for the class by totaling the number of tallies for A through E from each pair of students and comparing these numbers to the total number of spins. For example, if 85 out of 400 spins land on A, the experimental probability is 85/400. Ask students to estimate what common fraction 85/400 is close to (1/5) and to explain how the design of the spinner affects the results. Students may suggest that the areas of the sections of the circle are the same. Many students will also notice that the sections divide the circle into five equal pieces. Ask students questions about the number of times the spinner will land on section B when it is spun 40, 126, and 217 times. (About 8, 25, and 43) Although we can predict that 8 out of 40 spins will land on section B, for example, there might be quite a bit of variation when student pairs examine their own data.

You may want to ask students what other devices they have used that have equally likely outcomes when they are put in motion (e.g., coins when tossed, number cubes when rolled).

To help students interpret the distribution of spins for any pair of students, explain that they are going to make a line plot to show the distribution of spins for A, B, C, D, and E for spinner 1. If students have not worked extensively with line plots, you may wish to ask questions such as these: "In forty spins what is the greatest number of times the spinner could land on C?" (40) "What is the least number of spins that might land on C?" (0) "Do you think it is very likely that anyone will get either of those extreme values?" "Why, or why not?" (It is very unlikely but not impossible. Since there are five equal sections of the circle, it is likely that the spins will land on all five sections.) "Did anyone's spinner land on C zero or forty times?" Guide the students in making line plots that show graphically how many spins landed in sections A through E. Tell them to use their own data and place an X on the line plot to represent the number of times a spin landed in a particular section. While students are working, put the line plot below on the board.

A	B	C	D	E
	X			
	X		X	
	X	X	X	
	X	X	X	X
X	X	X	X	X
X	X	X	X	X
X	X	X	X	X
X	X	X	X	X
X	X	X	X	X
X	X	X	X	X

When everyone has recorded his or her data, ask students to describe both the line plot on the board and their own line plots. Some students will notice that on their line plots, the number of spins is fairly evenly distributed among the letters, whereas other line plots will show that one or two letters were spun more often. In the line plot on the board, section B is the most frequent result, but the numbers of spins for all the letters cluster around eight. Ask students to describe the shape of the data on their line plots and to conjecture how the design of the spinner is reflected in the line plot. Many students think that such a spinner will land equally on each of the five sections even when spun a mere five times. It is important for them to see that the distribution of spins levels off, or evens out, with more and more spins. Discuss how a large number of spins has a better predictive value than a small number of spins.

You can also construct a class line plot that shows the distribution of spins for one letter—say, D. Ask each pair of students to indicate how often their spinner landed on section D and plot the numbers on a class line plot. A possible class line plot is shown below.

1	2	3	4	5	6	7	8	9	10	11	12	13	14	15	16	17	18	19	20	25	26	27	28
						X	X																
				X		X	X		X	X													
X	X			X		X	X	X	X	X			X	X	X				X				X

The amount of variation among the students' line plots may surprise you. Variation is not unusual, but the median number of spins that land on D should be around eight. Ask questions about the center, the range, and the shape of the data. You may have an outlier in the data (see chapter 2), but most data points are likely to form a mound. Depending on time and interest, you may want to have students construct class line plots for the other outcomes (A, B, C, and E).

Turn the investigation next to a discussion of spinner 2. Examining this spinner, you will notice that the outcomes are not equally likely; the areas of the five sections are not identical. Have students collect

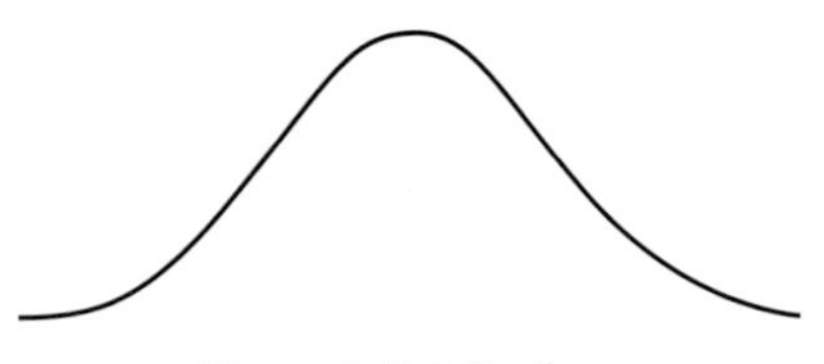

Normal distribution

A bell-shaped distribution is referred to in statistics as a normal distribution. In a normal distribution, the majority of occurrences are in the middle (hence the mound, or bell shape), with fewer and fewer occurrences as you move farther out on both sides.

data by spinning the spinner twenty times and recording the outcomes of their experiment on a chart on the board. Use their data to calculate the experimental probability for the class. Perhaps ask different pairs of students to come to the board and generate the experimental probabilities for the different sections, A through E, and to explain their reasoning to the class. Students often make sense of material when they hear it explained in a number of ways, so be sure to ask more than one pair of students for explanations. You also will be able to use this "reporting out" strategy as an assessment tool to gauge students' levels of understanding. As the class discusses the experimental probabilities, they will be able to conclude that landing on A, B, C, D, and E are not equally likely events. The theoretical probabilities for landing on each section are as follows: A—1/5, B—1/2, C—1/10, D—1/10, E—1/10. Theoretical probabilities can be calculated by analyzing the geometry of the circle. Experimental probabilities are based on experimental data. When large numbers of trials are conducted, experimental probabilities approximate theoretical probabilities.

Ask students to estimate the number of spins that would land on the different sections if the spinner is spun 40, 124, or 250 times. (About 8, 25, and 50 times for A; about 20, 62, and 125 for B; and about 4, 12, and 25 for each of C, D, and E) Also be sure to pose these questions: "Why are these numbers just estimates?" (The numbers are estimates because when we conduct an experiment, the spinner might not land on A exactly 50 out of 250 times. Also, if we spin only a small number of times, the results might be different.) "Why are the estimates different for some of the letters?" (They are different because the sections on spinner 2 are not the same size—the spinner will land on the biggest section, B, more often.)

You may want to have three pairs of students post their line plots for everyone to see and discuss.

Ask pairs of students to make a line plot from their tallies to show the number of spins for each letter. Suggest that after they complete their line plots, they compare them with someone else's. What do they notice? Although the overall shapes of the line plots will be similar (small amount on A, large spike on B, and C, D, and E with about the same number of spins but fewer than A), there will be variation among the line plots. For example, even though section E has a smaller area than section A, more spins could have landed on E than on A in 40 spins. However, when the data are combined for the whole class, showing results from 400 or more spins, section A should have about twice as many spins as section E.

Have students compare the line plots of spinner 1 and spinner 2. The line plot with equally likely outcomes (spinner 1) will probably be fairly level across the letters, whereas the line plot with unequal outcomes (spinner 2) will show different levels and noticeable peaks and valleys. Have students both describe the line plots and compare them with the spinners. Next, distribute copies of the blackline master "Matching Line Plots with Spinners" to students and explain that they will be matching line plots to spinners that possibly provided data for them. Remind them of the variation that can occur among spinner data and encourage them to look for patterns that may indicate how the spinners are divided. Suggest that they work with their partners but tell them that each student must be able to justify the match of a line plot with a spinner.

While students are working, circulate among them, posing questions and answering them. You may need to direct students' attention to certain relationships by asking questions such as "How can you tell from the line plot the number of sections into which the spinner is divided?" and "What might be true of a line plot of equally likely events?" When everyone has completed the handout, invite the class to discuss two or three of the problems. In each situation, ask students to explain why they believe that a certain line plot and spinner match. Ask other students if they agree or disagree with the first student's conclusion, and why. Typical responses for how they matched up a line plot and a spinner might include the following: "I noticed there were four letters on line plot *(a)* so I looked only at spinners with four sections. Then I noticed that one of the spinner letters (C) had lots more spins than the other three. This suggested that it matched with the last spinner *(h)*, since it has unequal sections and section C was much larger than the other three."

In the blackline master "Matching Line Plots with Spinners," students should match–

- *(a)* with *(h)*;
- *(b)* with *(g)*;
- *(c)* with *(e)*; and
- *(d)* with *(f)*.

Extend

This lesson can be extended by using the applet Probability Games on the accompanying CD-ROM. This applet includes activities called Preset Spinners and Make Your Own Spinners, which simulate spinning a spinner 1, 5, and 10 times. Students can select a spinner from the menu and predict what the data from multiple spins will look like when plotted on a tally frequency chart. They can then watch as the computer spins the spinner and electronically records the outcomes in a frequency table. A variety of spinners are available—some for which the outcomes are equally likely and some for which the outcomes are not equally likely. The applet includes an additional activity called Dice Sums, which students can also use to extend their understanding of probabilities when outcomes are equally likely and when they are not. Students can work with pairs of dice with preset numbers, or they make their own dice.

Assessment Ideas

One way to assess students' understanding is to ask them to compare in writing the expected outcomes from spinners 1 and 2 on the blackline master "Spin It." Even though students have discussed these spinners in detail in class, writing explanations is a complementary task and will provide you with insights into individuals' understanding. In addition, you can collect copies of the blackline master "Matching Line Plots with Spinners" as part of your assessment of the students. If many students had difficulty matching spinners with line plots or with answering the questions about their reasoning, review these tasks and consider incorporating more work with spinners and line plots into later probability lessons.

Where to Go Next in Instruction?

Have students design their own spinners, collect data using their spinners, and draw line plots of the data distribution. Make sure that they spin their spinners at least forty or fifty times. Some students may want to record the representations (pictures of their spinners and line plots of their results) on paper and then have everyone else in the class match the spinners and line plots, as in the blackline master "Matching Line Plots with Spinners."

For additional activities on how spinners can be used to gather probability data, see Mason and Jones (1994) on the CD-ROM.

Coins, spinners, and number cubes are used regularly in probability experiments, since predicting the outcome of experiments and then collecting data with these models is straightforward. These models are also used in games of chance. In the next activity, Is It Fair? students again gather data, but this time they also consider whether or not a game is fair. They are challenged to make sense of the fact that if a game is fair, then each player in the game should have an equal likelihood of winning.

Is It Fair?

Grades 4–5

This activity asks students to estimate and then determine the probability of a particular outcome of a simple game by playing the game and collecting data. Another purpose of the activity is to consider the meaning of fairness in a game. Students play a two-person game and determine the experimental probabilities of winning and losing for both players. They play again with different rules and again determine the probabilities. They decide whether or not the game is fair, and they write about the need for fairness in games. Finally, they investigate different approaches to make the game fair for both players.

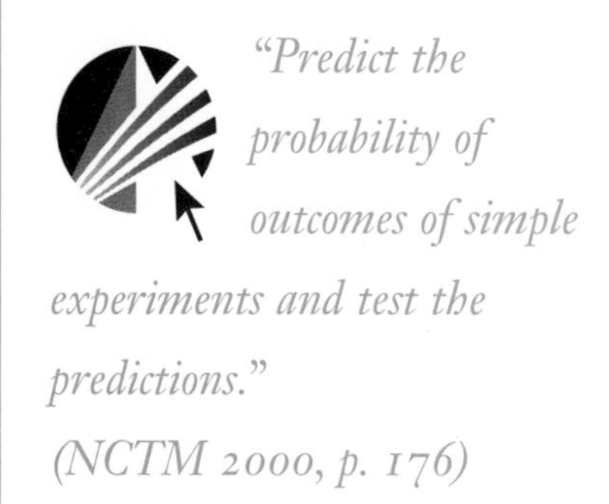

"Predict the probability of outcomes of simple experiments and test the predictions."
(NCTM 2000, p. 176)

Goals

Students will—

- understand that a fair game implies that there are equal probabilities of winning for all players;
- analyze the fairness of a game.

Prior Knowledge

- Students should be able to calculate experimental probabilities.
- Students should have an understanding of factors and multiples, prime numbers, and even and odd numbers.

Materials and Equipment

- Copies of the blackline masters "Number Cards" and "Rule Cards" for each group of two, three, or four students playing the game

pp. 117, 118

Classroom Environment

Students play games with one or more partners and collect data. A whole-class discussion follows the data collection.

Activity

Engage

"What does it mean to say that a game is fair or not fair?" Pose this question to your class at the start of this lesson. Many students' comments will reflect the everyday meaning of fairness, which is that something is equitable or just. Taking turns in a game and giving everyone the same-sized piece of cake at a birthday party are examples of equitable behavior that students might mention. What is meant in probability by the term *fair* is that there is no bias or favoritism toward one outcome or another. A fair game is one in which all players are equally likely to win or lose. The rules are written so that they don't favor one player over another. However, this doesn't mean that if one player wins the first round, the other player will win the next round. Remember that in games of chance that are fair, it is only after playing the game a large number of times that the distribution of wins evens out. Eliciting students' prior knowledge about fair games will enable you to understand what they do and do not know about fairness, and this will help

You should differentiate between games of chance and games of skill. A game of chance can be fair or not fair, but games of skill are a different story. In a game of skill, players' skills (or the lack thereof) affect the outcome; one player might win because of expertise in a particular area. There also are games whose outcomes depend on both chance and skill.

you address misconceptions later in the lesson. You may want to write students' views of fairness on the board or make lists of games that they believe to be fair or unfair. Do not worry if some students' statements are incorrect—you will return to the discussion of fairness later, and at that time, you can eliminate statements that students have come to see as untrue.

Students can explore whether a game of chance is fair or unfair by playing the game many times and comparing the number of wins for each player. If the game is unfair, one player will have more wins than the other. If the game is fair, the number of wins for both players will be approximately the same. However, sometimes when a game of chance is unfair, the data gathered from playing it numerous times are still very confusing—it may appear that the game is fair even though it isn't! In addition, if a game of chance is played only a few times, it is almost impossible to make a reliable judgment on its fairness. To determine the fairness of a game of chance completely, one has to analyze its rules and structure for bias. This can be done by comparing the number of ways in which each player can win.

Explore

Introduce your students to a little game, "Does the Number Fit Your Rule?" The goal is to win as many points as possible in each round, and a player wins a point when a number card is turned over that fits the rule card that he or she continues to hold throughout an entire game. Two, three, or four students can play.

The game uses the cards that appear in the blackline masters "Number Cards and Rule Cards." Make copies of these blackline masters for each group of students that will be playing the game. Cut out the cards in advance and give each group a set of each type of cards, or distribute a sheet of number cards and a sheet of rule cards to each group and ask the students to cut them out. Direct the students to shuffle each set of cards and put them face down in two piles.

Taking turns, the players should each draw one of the rule cards. Then, one after another, the players should turn over the number cards. As a number card is displayed, each player must decide if the number "fits" his or her rule. If it satisfies the rule on a player's card, he or she wins a point. For example, if player A has the rule card "multiples of 2" and player B has the card "multiples of 5" and a 20 is turned over, then both players get 1 point, since 20 is a multiple of both 2 and 5. If a 15 is turned over, then only player B gets a point, since 15 is a multiple of 5 but not of 2. If a 3 is turned over, neither player gets a point, since 3 is not a multiple of 2 or 5. Turning over ten number cards and evaluating them against the same rule card makes a round. After playing ten number cards, players record their scores for that round. They then play two more rounds using the same rule card and keeping a cumulative total of the number of points. The winner of the game is the player with the highest total score for the three rounds.

You may want to review the term *multiple* by listing the first few multiples of a number such as 3 (e.g., 3, 6, 9, 12, 15, 18, 21, ...). Then give students a chance to list the multiples of 2 and the multiples of 5 before turning over number cards.

Model the game by playing with your class. Use the rule "multiples of 2" for yourself and the rule "multiples of 5" for the class. Before playing, ask the students if they think this game, with these particular rules, is fair or unfair. That is, do both players have the same chance of winning? Play two rounds with the class and again ask the students if they think the game is fair. Answers will vary depending on the outcomes of the first two rounds, but responses might include "It's hard to tell after only two rounds" and "It seems fair, since you got a lot of points in one round and we got a lot of points in the other round." Play the final round with the class and again discuss whether the game is fair. Next, suggest that students play the game with a partner.

This game is not fair, since out of the thirty number cards half are multiples of 2 (all the even numbers) but only six are multiples of 5 (5, 10, 15, 20, 25, 30). The chance of winning with multiples of 2 is 1/2, whereas the chance of winning with multiples of 5 is 1/5.

When students finish playing the game with these particular rules, again open up the discussion of whether the game is fair or not fair. Pose the question, "Which rule card would you like to have?" Focus on the fact that fairness in probability means that one player does not have an advantage over the other player. You may want to examine data from different pairs of students to highlight the unfairness of the game. A table like the one below is a sample of a game played by two students. These results are just one possibility out of many outcomes.

	"Multiples of 2" pts.	"Multiples of 5" pts.	Number of Cards
Round 1	6	3	10
Round 2	4	1	10
Round 3	5	2	10
Total	15	6	30

(*Note:* The number of cards turned over each time the game is played is ten, but this does not mean that the scores of the two players will add to 10. Some number cards do not result in points for the rule-card holder.) Other student pairs may have different scores for each round and different totals.

When students agree that it appears that this game using the multiples of 2 and multiples of 5 rules is not fair, suggest that one way to verify their suspicion is to analyze all the outcomes. Direct students to talk in groups of two to four to brainstorm why the two rules favor the multiples of 2. Expect students to mention that "there are more multiples of 2 in the number card deck than there are multiples of 5" or that "there are only six numbers in the deck that are multiples of 5, but there are fifteen cards that are multiples of 2!" You may wish to compare the number of cards that satisfy the rule to the total number of cards in the deck (e.g., 15/30 for multiples of 2 and 6/30 for multiples of 5).

Extend

There are many ways to extend this lesson. Suggest that students revisit the game that they have played so far (multiples of 2 versus multiples of 5) and change the rules so that the game will be fair rather than unfair. Students may use trial and error to change a rule and then collect data to gather information about the fairness of the games with their new rule. Students should be asked to explain how they have made the game fair. They might make unfair games, such as the "multiples of 2 and multiples of 5" game, fair by changing a rule, such as "multiples of 2" plays against "odd numbers." Another way to make it fair is to adjust the rule for one player so that "multiples of 5" becomes "multiples of 5 or multiples of 3." Students also can make games fair by changing the point values that each player receives. For example, the "multiples of 2"

See Wiest and Quinn (1999) on the CD-ROM for a description of a more complex game that helps students explore probability.

player could receive 2 points every time his numbers were drawn, and the "multiples of 5" player could receive 5 points for each of her numbers. Students will come up with additional ideas as well.

In addition, have students play the game using the other rule cards, each time not only collecting some data but also analyzing outcomes. Or have students make up their own rules for the game. Encourage them to use their knowledge of factors, multiples, prime and composite numbers, even and odd numbers, and place value to think up clues that will include some but not all of the numbers in the deck (blank rule cards can be found on the blackline master entitled "Rule Cards").

The game suggested in "Assessment Ideas" is fair. There are ten multiples of 3 in the deck (3, 6, 9, 12, 15, 18, 21, 24, 27, and 30), and there are ten prime numbers (2, 3, 5, 7, 11, 13, 17, 19, 23, and 29).

Assessment Ideas

Have students play another game where one player has the rule "multiples of 3" and the other player has the rule "prime numbers." Have them predict whether or not this game is fair, play the game ten times to get a sense of its fairness, and then analyze the game to determine if it really is fair or not. Have student pairs work on this game together but write up their results individually.

Where to Go Next in Instruction?

If you wish to expose students to other situations involving fairness, you can have them play other games that involve chips, number cubes, or coins and analyze them for fairness. It is important that students collect data by playing a game many times and that they use these data to help them determine the fairness of the game.

The activities in this chapter encourage students to focus on ideas regarding probability. Students learn that we can quantify the likelihood of an event using a value from 0 through 1 and that descriptors such as *impossible* and *likely* can be used to describe the probability of an event. Students expand on what they have learned about probability by conducting different types of simple experiments, gathering and analyzing data, and making predictions about the outcomes. Finally, they go on to predict the probability of simple experiments while they investigate the fairness of simple games.

GRADES 3–5

DATA ANALYSIS *and* PROBABILITY

Looking Back and Looking Ahead

In grades 3 through 5, instruction in data analysis and probability continues to build on students' experiences with organizing data, representing data, and describing data sets in kindergarten through grade 2. In the earlier grades, students often represent data using concrete objects, pictures, and simple graphs. They tend to analyze data by looking at individual pieces of data and by identifying which value is the "most." In grades 3–5, students' analyses of data go beyond these activities. Students in these grades need to have many experiences describing and comparing sets of data, examining patterns and trends, and noticing how data are distributed.

One of the goals of instruction at this level is for students to become familiar with a variety of representations, such as tables, line plots, bar graphs, and line graphs. It is not an easy task to decide on an appropriate graphical representation for a data set. Students must consider the meaning of the horizontal and vertical axes, the size of the scales, and how categorical and numerical data are shown on these axes. They need to discuss the reasons why certain numerical data (like temperature during a day) can be represented on a line graph, but other numerical data (like the numbers of books in the classrooms in a school) cannot. Temperature values from sunrise to sunset can be represented on a line graph because these numerical data are continuous (i.e., for any two values we can find one between them). Other numerical data, such as the numbers of book titles in the classrooms in a school, are discrete (i.e., there is no such thing as 3.5 books). In chapter 1, students not only construct graphs but also compare the different representations. It is through the comparison process that students start to evaluate how well certain aspects of the data are shown with each of the representations. By navigating among the different representations, students also learn to connect information represented both in a table and in a graph.

Another major goal of instruction in grades 3–5 is to have students start to grapple with the measures of center of a data set. It is in this grade band that students begin to learn to analyze, summarize, and describe an entire data set, first by learning how to examine the shape of the data and then by learning concepts of central tendency. Mean, median, and mode—the common measures of central tendency, or measures of center—are summary statistics. Summary statistics help us describe and compare data sets. Students need to understand the meaning of these measures and how to determine them. Friel (1998, p. 209) states that "the importance of students' having strong conceptual understandings of these concepts cannot be overstressed; the value in their use comes from understanding what these measures 'tell' about data." In grades 3–5, students' learning about measures of center needs to build on their informal understandings of concepts such as "the most," "the middle," and "the typical," and should emphasize the concepts of mode and median. Only after students have a firm understanding of mode and median should they move to explorations of mean.

The concept of the median is perhaps most important for students in grades 3–5 to comprehend. Friel suggests that teachers have students develop the idea of median with small data sets before working with large sets of data. Students should experiment with determining the median when the data are presented in unorganized lists, in line plots, and in graphs.

One way to build students' understanding of the measures of center is to ask them to decide which measure(s) to use to describe what is typical about a data set. For example, if data are categorical (e.g., favorite ice-cream flavors), the median and mean have no descriptive value. The mode, however, provides useful information by indicating that vanilla is the most popular flavor (see the illustration in the margin). Other situations that involve numerical data are perhaps better described using the mean and median as well as the range. Students need to be given opportunities to analyze data sets and should be encouraged to justify their decisions in answering questions on measures of center. For example, the line plot below shows the twelve students' sales of candy to earn money for a field trip.

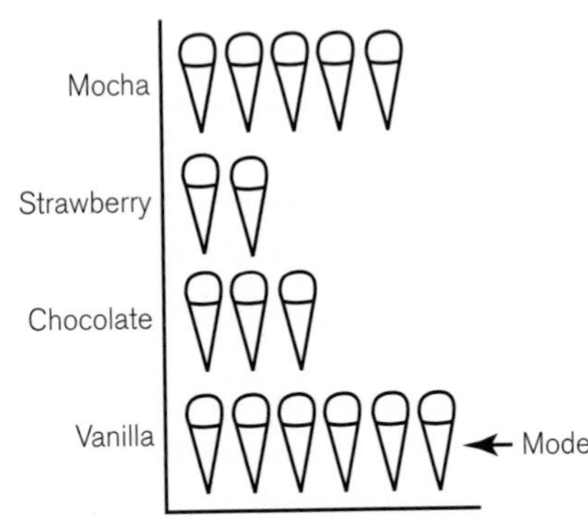

		X						
X		X						
X	X	X						X
X	X	X				X		X
$20	$30	$40	$50	$60	$70	$80	$90	$100

Students must recognize the fact that although the mean, or arithmetic average, is $46.67, most students did not sell that much. In fact, the median, or middle value, of this data set is $40, and nine out of twelve sold $40 or less! Some students will insist that the mode gives the "average" amount of sales, whereas other students will insist it is important to present both the median and the mean to fully describe the situation.

It is in grades 3–5 that students first start to develop and evaluate inferences that are based on data. Consider the following situation from Garfield and Gal (1999, p. 214):

> For one month, 500 elementary school students kept a daily record of the hours they spent watching television. The average number of hours per week spent watching television was twenty-eight. The researchers conducting the study also obtained report cards for each of the students. They found that the students who did well in school spent less time watching television than those students who did poorly.

It is important to emphasize that even a strong relationship among data sets is not enough to establish causality.

Students in one class debated whether or not the study above proved that watching television causes lower grades. They used this situation to pose questions and develop conjectures, consider alternative explanations, and list what additional information was needed in order to draw conclusions. They concluded that "even though students who did well watched less television, this doesn't necessarily mean that watching television hurts school performance."

Students in grades 3–5 also are exposed to the idea that the data sets they are examining are samples of larger populations. They start to make sense of samples by taking many samples from the same population—surveying all the fourth-grade classrooms in their school or comparing data from their school with data from other schools across the state. Under these circumstances, they start to consider issues that they will investigate more thoroughly in grades 6–8—sample size, sample bias, and random-sampling techniques.

In grades K–2, students are informally exposed to ideas about the likelihood of events. It is in grades 3–5 that students learn more formally about probability and that the likelihood of an event can be quantified. One of the obstacles that teachers will have to confront is that students bring with them many inconsistent beliefs and misconceptions about the way things perform in probabilistic situations. For example, when students consider the likelihood of a coin landing heads up, they often think back to the few times they may have flipped a coin. They may make conjectures about the outcome from their own experiences (e.g., "It landed on heads most of the time"). Thus, they may assume that most coins land heads up. For students to make better predictions about the probability of outcomes, it is essential that they engage in data collection. When students investigate the likelihood of heads by tossing dozens of coins, they are faced with results that do not correspond with their preconceived notions. The cycle of prediction, experimentation, data collection, and analysis of results forms the basis for instruction at this level.

See Jones et al. (1999) on the CD-ROM for a description of students' probabilistic reasoning.

A foundation for understanding data and probability starts in grades K–2 and continues through middle school to high school statistics and probability experiences. Students in grades 3–5 should encounter the ideas presented in this book and briefly summarized here. They need instructional experiences that engage them in investigating problems and situations in data analysis and probability by conducting experiments, collecting and displaying data, and, finally, analyzing the data in light of the initial questions and problems.

Grades 3–5

Data Analysis *and* Probability

Appendix

Blackline Masters and Solutions

Determining a Purpose for a Data Investigation

What is the relationship between the sample questions and the purpose of a data investigation?

To describe or summarize what you have learned from a set of data:

How often does ...?

How often does *the average fourth grader wash his or her hands in a day?*

How many ...?

How many *pieces of paper were on the floor in each classroom?*

How much ...?

How much *did students in the class weigh when they were born?*

How long ...?

How long *do hurricanes last?*

Which ...?

Which *class read the most books last month?*

Determining a Purpose for a Data Investigation

(continued)

To determine preferences and opinions from a set of data:

What is your favorite …?

What is your favorite *music group?*

Which is the tastiest …?

Which is the tastiest *brand of chocolate?*

With whom would you …?

With whom would you *want to spend time?*

What would you do to …?

What would you do to *improve your school?*

What … do you value …?

What *traits* do you value *in a friend?*

What is the best method for …?

What is the best method for *getting dressed for school?*

Determining a Purpose for a Data Investigation

(continued)

To compare and contrast two or more sets of data:

What are the similarities and differences between ...?

What are the similarities and differences between *third- and fifth-grade students' favorite rides at an amusement park*?

Does the ... differ between ...?

Does the *number of chores expected by parents* differ between *third and fourth graders*?

What is the relationship between ...?

What is the relationship between *the temperature of water and the amount of time it takes to dissolve a cube of sugar*?

Determining a Purpose for a Data Investigation

(continued)

To generalize and make predictions from a set of data:

What is the typical …?

What is the typical *type of book read by students in the class?*

Can you predict …?

Can you predict *your neck measurement from your wrist measurement?*

How would you …?

How would you *react to going to school year round?*

Is there a trend …?

Is there a trend *between the months of the year and the number of student absences?*

Getting Ready

Name ______________________________

? ? ? ? ? ? ? ? ? ? ? ? ? ? ?

1. Purpose: To summarize what was learned from a set of data.

Question, please? ______________________________

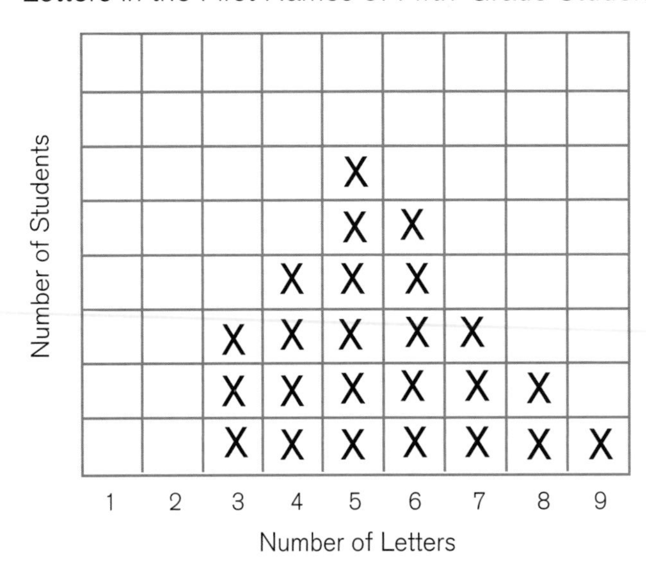

2. Purpose: To determine preferences and opinions from a set of data.

Question, please? ______________________________

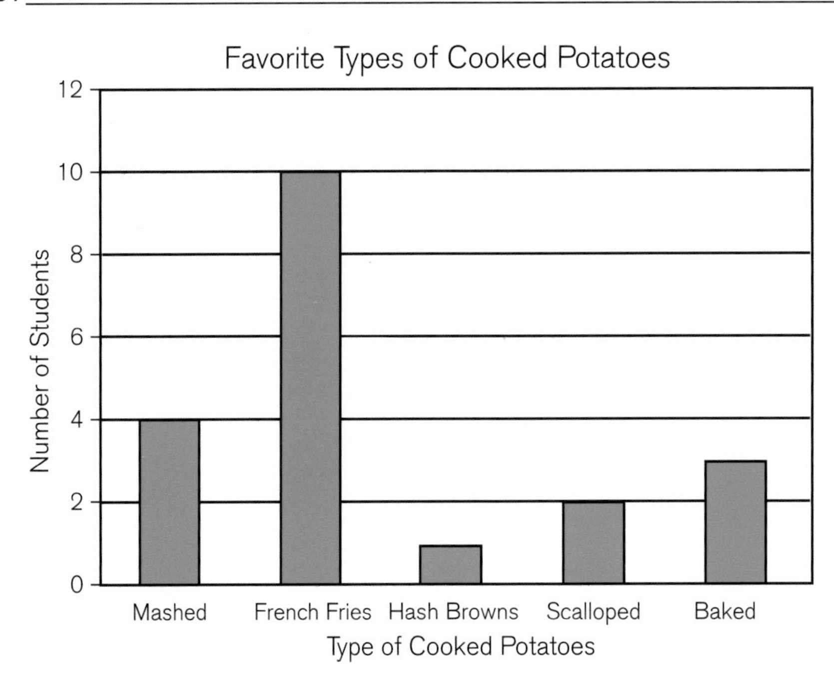

Getting Ready (continued)

Name ______________________________

? ? ? ? ? ? ? ? ? ? ? ? ? ? ?

3. Purpose: To compare and contrast two or more sets of data.

Question, please? ____

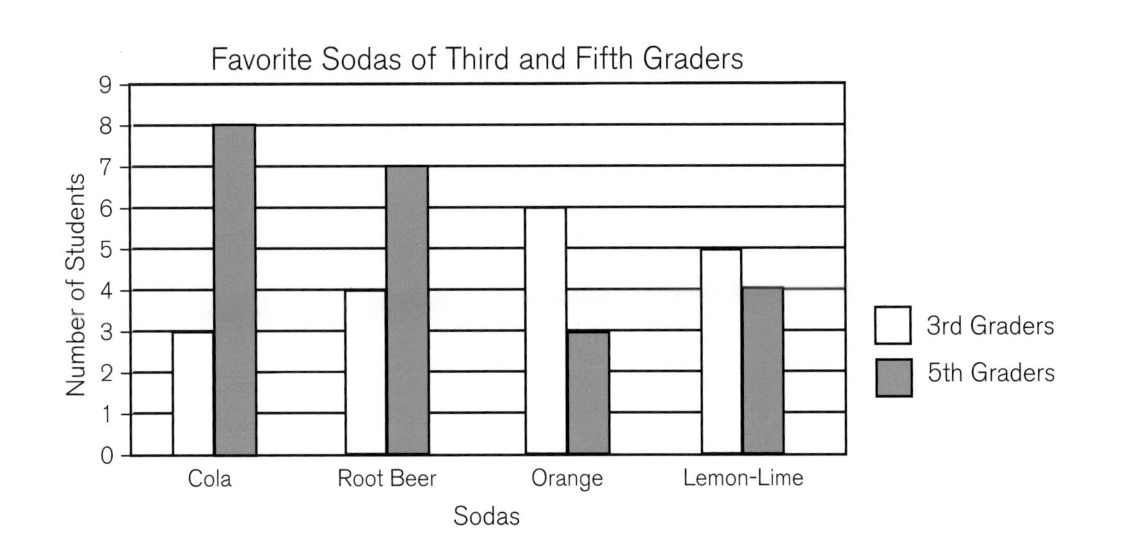

4. Purpose: To generalize or make predictions from a set of data.

Question, please? ____

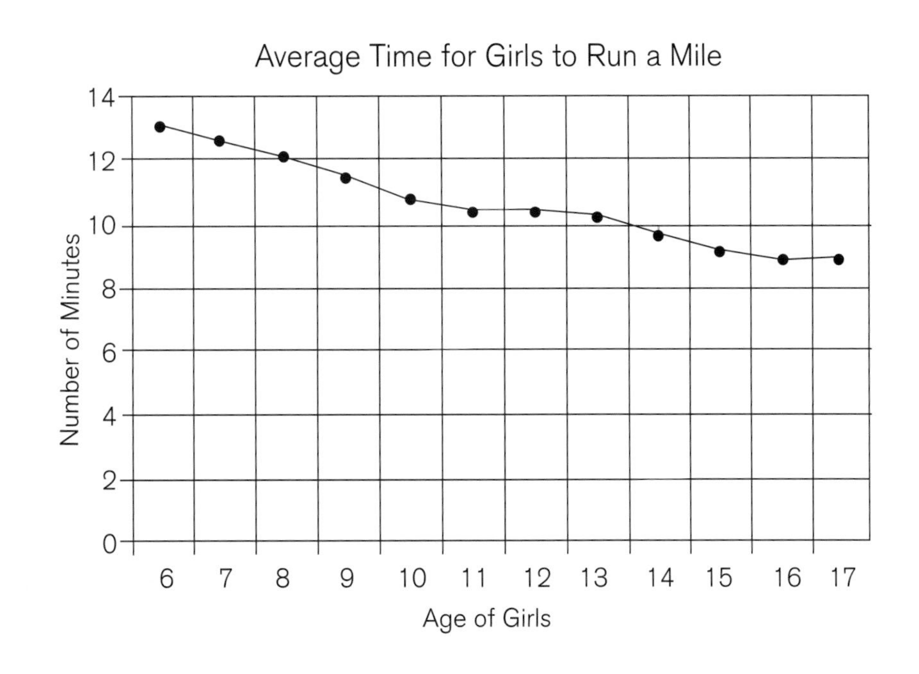

A Question to Investigate

Name ______________________________

Formulating Questions for Data Investigation and Analysis

First, select a topic for your data investigation from the subjects, themes, and areas of interest in the list:

Amusement parks
Archaeology
Body measurements
Books/literature
Childhood diseases
Clothing/fashion
Communities
Community helpers
Computers
Dinosaurs
Environment
Fairness
Five senses
Food

Friendship
Games
Geography/travel
Health habits
Highway safety
Holidays
Jobs/earning money
Littering
Movies
Music/musical groups
Mysteries
National disasters
Nutrition habits

Organizations
Outer space
Planets
Pollution
Science experiments
Science fiction
Seasons/weather
Second-hand smoke
Shopping
Sports
TV shows
Vacations
Waste

Selected subject/theme/interest: ______________________________

Selected topic: ______________________________

A Question to Investigate (continued)

Name ______________________________

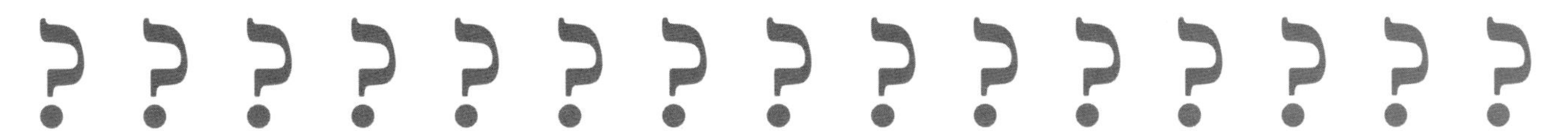

Second, select a purpose for your data investigation and use one of the question stems to help you formulate your question. You may use a question stem that is different from the ones listed below.

To describe or summarize what you have learned from a set of data

How often does …?
How many …?
How much …?
How long …?
Which …?

To determine preferences and opinions from a set of data

What is your favorite …?
Which is the tastiest…?
With whom would you …?
What would you do to …?
What … do you value …?
What is the best method for …?

To compare and contrast two or more sets of data

What are the similarities and differences between …?
Does the … differ between …?
What is the relationship between …?

To generalize and make predictions from a set of data

What is the typical …?
Can you predict …?
How would you …?
Is there a trend …?

Third, formulate your question on the basis of the topic identified above:______________________

____

____

Data Sets

Students' Favorite Type of Pancakes

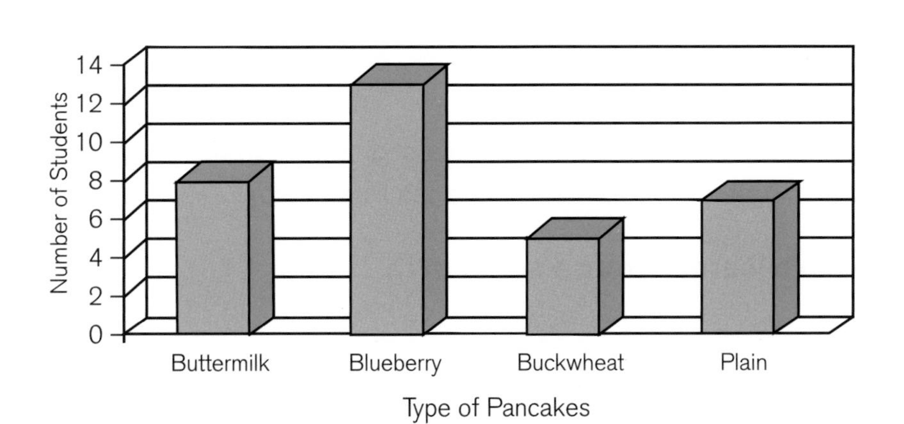

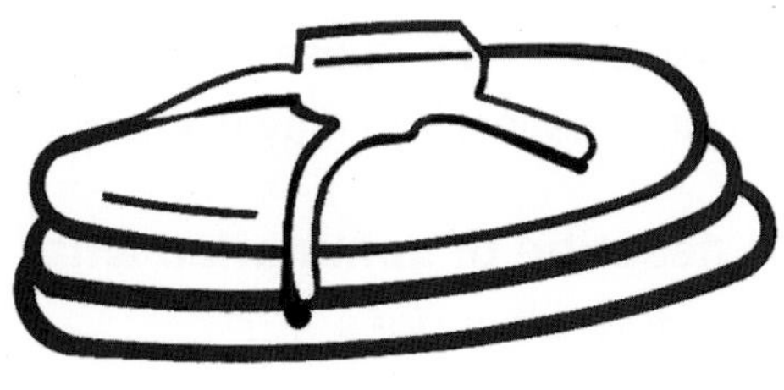

Result of Student Elections for Class President

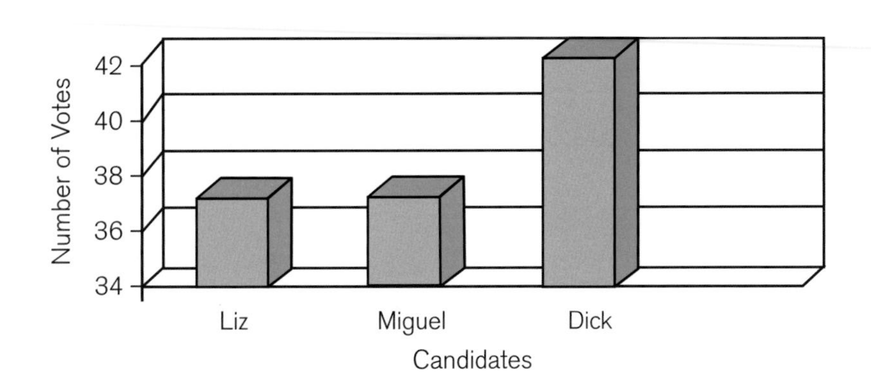

Dog Database

Dog's Name	Gender	Weight (kg)	Body Length (cm)	Tail Length (cm)	Eye Color
Skipper	M	30	58	13	Brown
Molly	F	20	45	25	Blue
Buddy	M	37	88	22	Brown
Jack	M	16	35	20	Green
Patty	F	38	65	18	Gray
Precious	F	4	15	8	Blue
Sam	M	50	75	30	Brown

Data Sets (continued)

Low Temperatures (in Degrees Farenheit)
Recorded during the Month of October

3	0, 2, 3, 3, 5, 7, 7
4	1, 4, 5, 6, 7, 8, 9
5	0, 2, 5, 6, 7, 8, 9, 9, 9
6	3, 4, 5, 5, 6, 7, 8, 8,

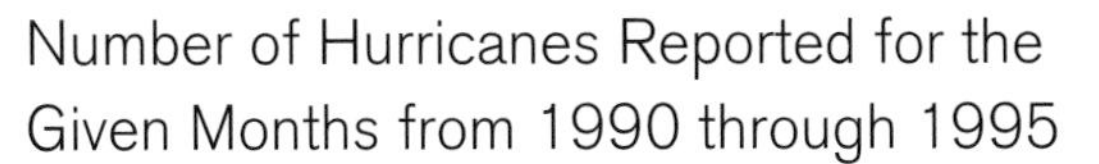
Number of Hurricanes Reported for the
Given Months from 1990 through 1995

Month	Number
June	11
July	17
August	43
September	67
October	25
November	9

What's My Method?—Descriptions

Data-Gathering Methods	Description
	*Observation* (O) is a method of collecting data in which a person records what he or she sees (e.g., records the different types of birds seen in the backyard) or what he or she does (e.g., records the number of books read in a month).
	*Surveys and questionnaires* (SQ) are similar tools and are used to ask people a question or set of questions that is the same for everyone. People often write their answers or select from a set or sets of possible answers.
	*Experiments* (E) entail collecting data under particular test conditions and often use special tools or techniques for measuring results, such as thermometers, rulers, calculator-based laboratories (CBLs), and probes.
	*Interviews* (I) are formal face-to-face meetings in which interviewers pose questions to people who have agreed to answer questions.
	*Polls* (P) are collections of votes or opinions, as in elections or public surveys, in which there are usually clear choices.

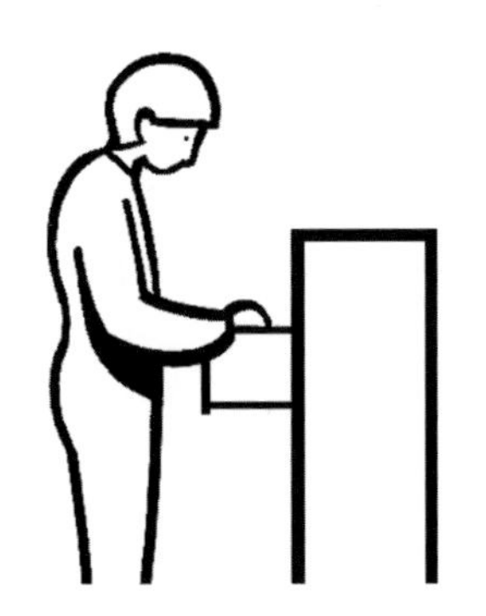

An examinination of past records (R) is a process of finding existing data instead of collecting new data. Looking up the number of baseball games won by the home team last year is an example of this method of data collection.

Searches of the Internet, library, or other resources (S) are ways of locating other existing information for examination in a data investigation.

A simulation (SI) is the act or process of imitating actual conditions through test conditions—for example, flipping a coin to represent the birth of girls (heads) versus boys (tails).

What's My Method?—Explorations

Name ______________________________

Identify methods that you would use to collect data to answer the questions posed below. Write the letter(s) for the method, selecting from:

O = observations
P = polls
R = examinations of past records
E = experiments
SQ = surveys and questionnaires
I = interviews
SI = simulations
S = searches of the Internet, library, or other resources

Question	Method
1. How many hours of television does the typical student watch in a week?	
2. How many jellybeans of each color would you expect to find in a sixteen-ounce bag?	
3. What are the differences between the annual rainfall in your hometown and the annual rainfall in each of three cities you most want to visit?	
4. How many hours do students listen to CDs in a week?	
5. How high can the typical student in your grade reach without jumping?	
6. Who will be elected as class officers in the next election?	
7. How many drops of water can you drop on the surface of a nickel before they spill over the edge?	
8. What would you predict to be the greatest differences in height between the shortest and the tallest students in any classroom in the school?	
9. What was the population of your state or province at the beginning of each decade since 1900?	
10. What was the total school enrollment for each of the years your school has been open?	
11. What is today's temperature in 10 capitals around the world?	
12. What are the three most favorite activities of the members of your family?	
13. What color appears most often in 20 spins of a four-color spinner?	
14. How far away do students in your class live from school?	

Summer Olympics 2000

Name ______________________________

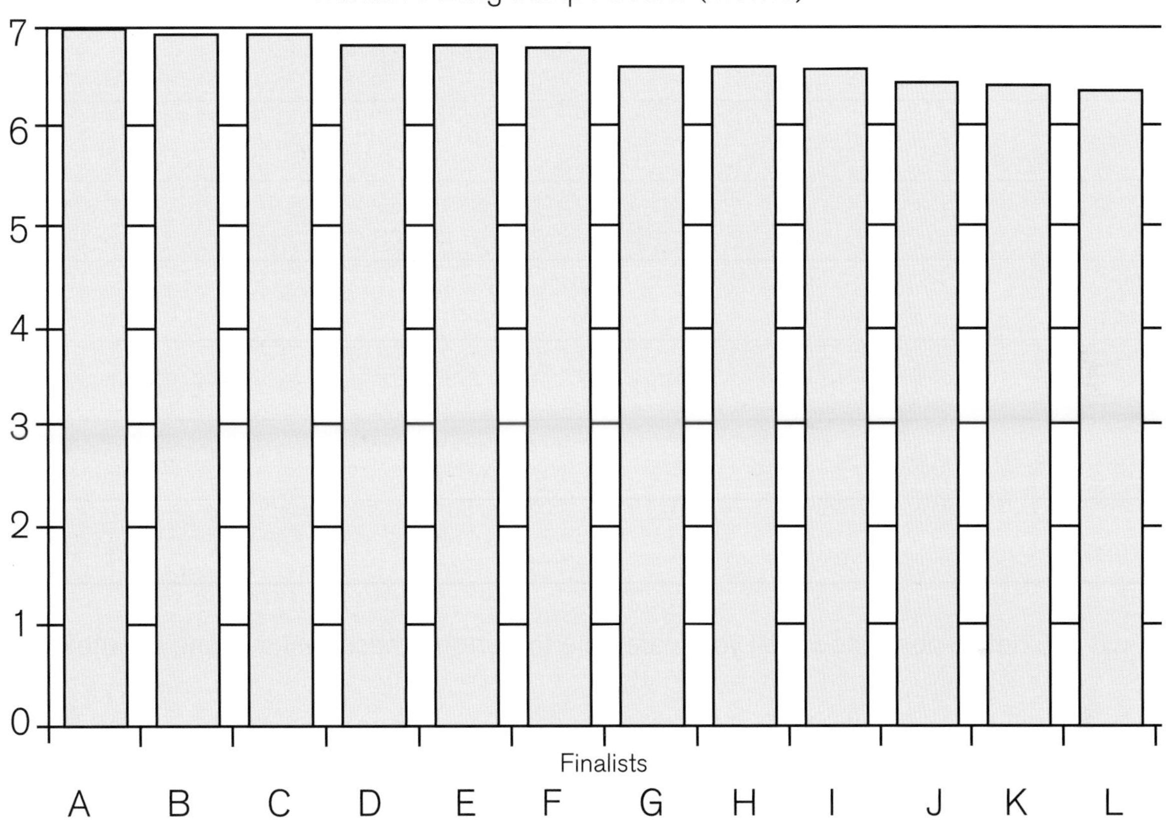

How Long Is One Minute?

Name ______________________________

Task	Prediction	Actual

1. What did you notice about how well you were able to perform these tasks in one minute?

2. Were you surprised by any of your results? Explain.

3. Brainstorm some other tasks you think you could repeat many times in one minute.

How Many Stars?–Another Class

Name ______________________________

The line plot below shows the data that a class at another school collected on how many stars each student could draw in one minute.

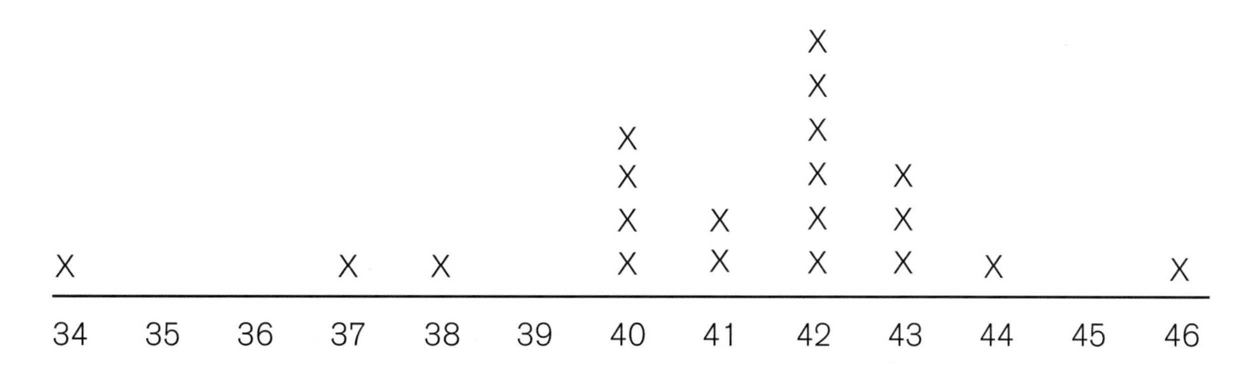

1. Examine the line plot. With your group, discuss these questions:

   - How are the data spread out?
   - Are there any clusters? Any gaps?
   - Are there any unusual data values?
   - How would you describe the typical star-drawing rate of this class?

   Write three or four statements about the data from this class.

2. Your class will be conducting an experiment to see how many stars each of your classmates can draw in one minute. How many stars do you predict *you* can draw in one minute? Explain your prediction.

3. What do you predict a line plot of your class's data will look like? Draw your predicted line plot. Explain your prediction.

How Much Sleep Do You Get?

Name ______________________________

Time you went to bed: ________

Time you awoke: ________

1. Determine the time you slept. Round your answer to the nearest quarter hour. Show your work below.

   Amount of time I slept: ______

2. Do you think this amount of sleep is typical for you? Explain.

3. How do you think the amount of sleep you got last night will compare with the amounts that your classmates got? Explain.

How Much Sleep Do Children Typically Get?

How much sleep do children actually get? How much do they need?

Age	Actual Hours of Sleep	Hours of Sleep Recommended
9 and under	9 1/4	10 1/4
10	9	9 3/4
11	9	9 3/4
12	8 1/4	9 1/4
13 and up	7 1/4	8 1/2

Source: Zillions magazine, May/June 1998

Women's Soccer Results

Name ______________________________

In the last seven games of the 2000 season, the USA Women's National Soccer Team had an average score of 2 goals per game.

1. Create two different line plots with data values for these seven games. Remember that the mean number of goals for these seven games is 2. Your data must have a mean of 2. Below each line plot explain how you know that it is a possible data set.

2. Create four additional sets of data values for the seven soccer games and display them on line plots that you label **a** through *d*. Remember that each data set must have a mean of 2. In addition, for each line plot you make, you must satisfy an added requirement:

   a. Only one game had 2 goals.
   b. Exactly two games had 4 goals.
   c. One game had an amazing 7 goals!
   d. The median of the data set is 3.

Women's Soccer Results (continued)

Name ______________________________

3. Describe the strategies you used to create the four line plots, *a* through *d*.

4. What if the mean score of the last seven games were 4 instead of 2? Create two different data sets for situations in which the mean is 4. Then explain the reasoning you used to create these data sets.

The Foot, the Whole Foot, and Nothing but the Foot—Group Data

Name ________________________________

Line Plot of Group (Sample) Data

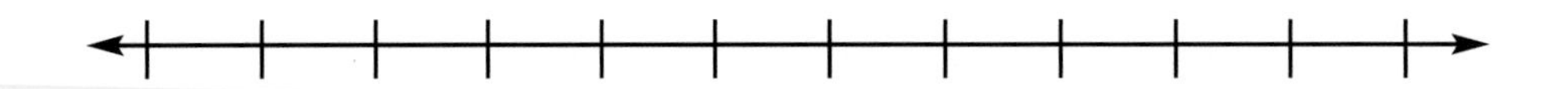

Foot Length (centimeters)

1. What does the line plot show? ________________________________

2. What is the shortest foot length in centimeters in this sample? ________________

3. What is the longest foot length in centimeters in this sample? ________________

4. What is the range of the foot lengths in this sample? ________________

5. What is the most common foot length in this sample, or the mode for this set of data? ________

6. What is the median foot length in this sample? ________________

7. Would you predict that this group's line plot would be representative of the population—that is, would it be similar to a line plot of the entire class's foot lengths? ________________

   Why, or why not? ________________________________

The Foot, the Whole Foot, and Nothing but the Foot–Class Data

Name ______________________________

Line Plot of Class (Population) Data

Foot Length (centimeters)

1. What does the line plot show? ______________________________

2. What is the shortest foot length in centimeters in the whole class? ______________________________

3. What is the longest foot length in centimeters in the whole class? ______________________________

4. What is the range of the foot lengths in the whole class? ______________________________

5. What is the most common foot length in the whole class, or the mode for this set of data? __________

____

6. What is the median foot length in the whole class? ______________________________

7. What is the mean foot length in the whole class? ______________________________

Can You Catch Up?

Name ______________________________

Length of the Ketchup Flow in Centimeters	Ketchup Temperature		
	Cold	Room	Hot
Prediction			
Actual			

Bar Graph of Predicted and Actual Length of Ketchup Flow for Three Temperatures

Length in Centimeters	Cold Prediction	Cold Actual		Room Prediction	Room Actual		Hot Prediction	Hot Actual
19								
18								
17								
16								
15								
14								
13								
12								
11								
10								
9								
8								
7								
6								
5								
4								
3								
2								
1								
0								

Write a question that could be answered by an analysis of the graph. ______________________________

____

____

Chores—How Many Hours a Week Are Typical?

STUDENT SURVEY

Name:______________________________________ Class ________________________________

1. Place a check mark next to the chores that you do each week.
2. Calculate the number of minutes that you spend performing each chore during the week and record it in the column titled Minutes a Week.
3. Add the numbers in the Minutes a Week column to determine the total number of minutes a week spent on chores and record this on the line given.

✔	Chore	Minutes a Week
	Caring for a pet	
	Caring for others (sisters, brothers, etc.)	
	Kitchen duties (setting the table, emptying dishwasher, etc.)	
	Cleaning your room	
	Household chores (vacuuming, dusting, etc.)	
	Yard work	
	Others (list them):	

TOTAL NUMBER OF MINUTES: ____________________

Do Not Write Below the Line

Students conducting the survey should complete the following steps:

1. Look at the total number of minutes that the student above spends on chores in a week. Use your calculator to convert this number of minutes to the number of hours, in decimal form, by dividing by 60 (60 minutes to an hour).
2. Round the quotient to the nearest tenth to identify the number of hours that this student spends performing chores (e.g., 370 minutes = 6.2 hours).

HOURS IN DECIMAL FORM: ____________________

Stem-and-Leaf Plot of the Group's Sample Data

Name ______________________________

Hours	Parts of an Hour (Tenths)

Sample versus Population–How Do They Compare?

Name ____________________

Parts of an Hour (Tenths)	Hours	Parts of an Hour (Tenths)

Paper Toss Recording Sheet

Name ______________________________

Student Data Sheet

Distance	Number of Tosses In	Number of Tosses Missed	Number of Total Tosses
3 cm			
150 cm			
250 cm			
350 cm			
6 m (eyes closed)			

Class Data Sheet

Distance	Number of Tosses In	Number of Tosses Missed	Number of Total Tosses
3 cm			
150 cm			
250 cm			
350 cm			
6 m (eyes closed)			

Impossible — 50-50 — Certain

0 — 1/2 — 1

Spin It

Name ______________________________

Spinner #1

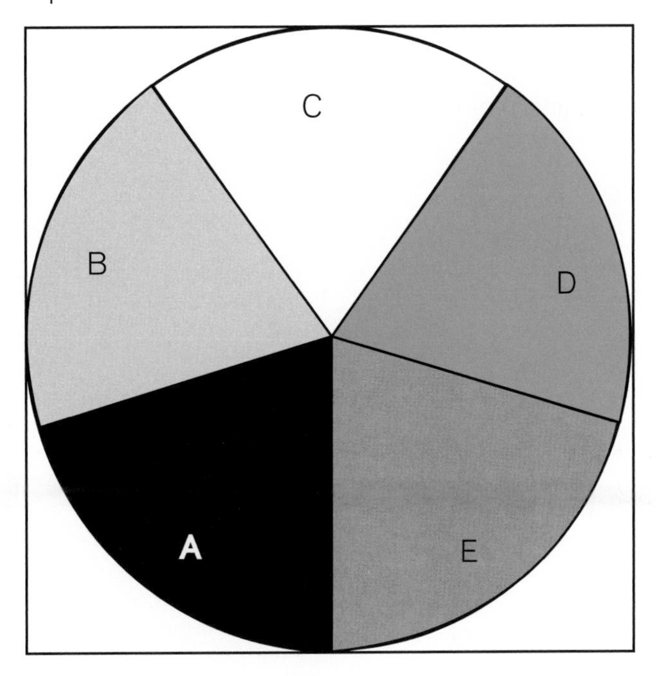

Spinner #2

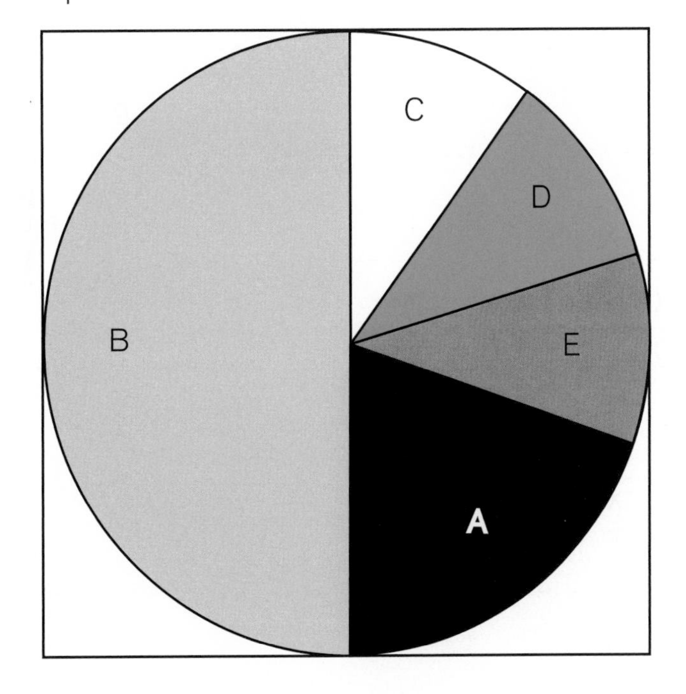

Matching Line Plots with Spinners

Name ______________________________

Match each line plot with a spinner. When you are finished, answer these questions:

1. Explain how you knew which spinner matched line plot *b*.

2. Explain how you knew which line plot matched spinner *h*.

a.

```
      X
      X
      X
      X
      X
      X
      X
      X  X
      X  X
X  X  X  X
----------
A  B  C  D
```

b.

```
X  X     X  X
X  X  X  X  X
X  X  X  X  X  X
----------------
A  B  C  D  E  F
```

c.

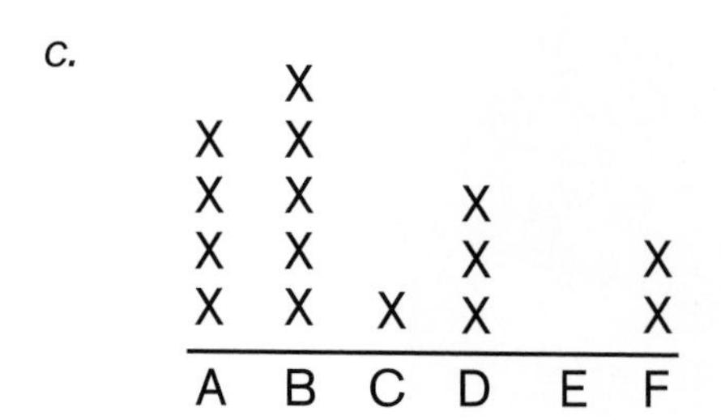

d.

```
         X
X        X
X  X  X  X
X  X  X  X
X  X  X  X
----------
A  B  C  D
```

e.

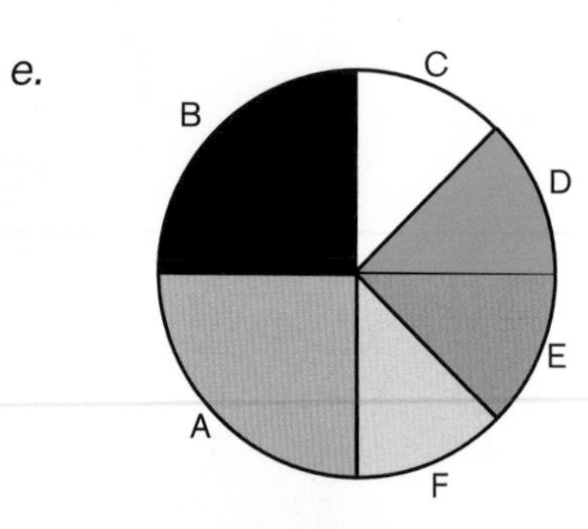

f.

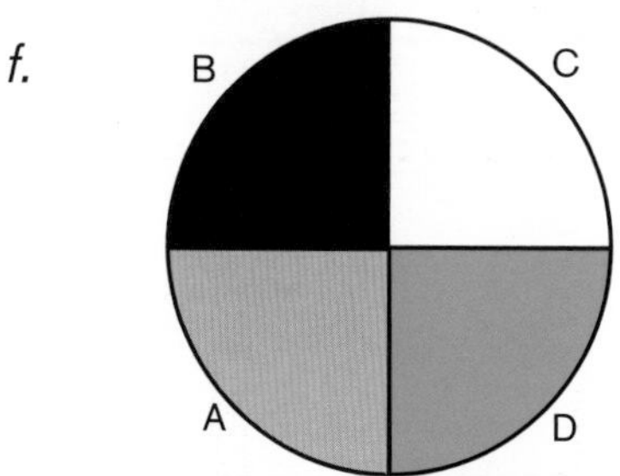

g.

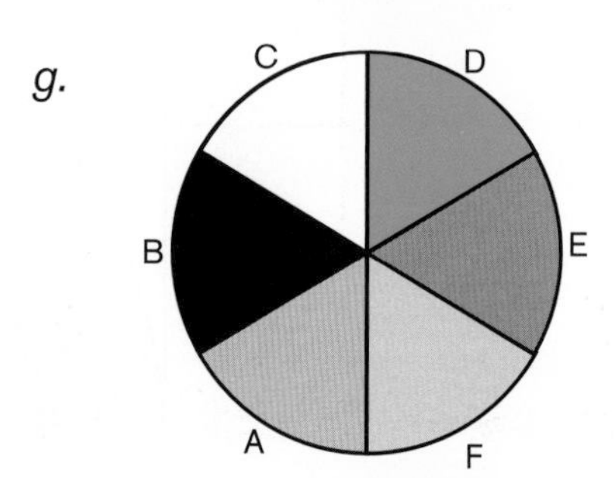

h.

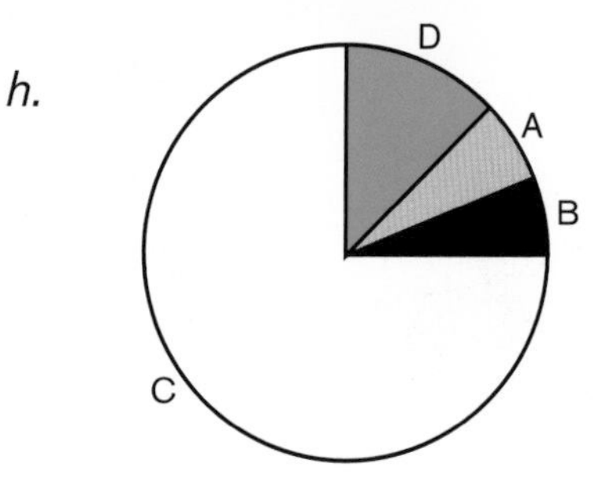

Number Cards

1	2	3	4	5
6	7	8	9	10
11	12	13	14	15
16	17	18	19	20
21	22	23	24	25
26	27	28	29	30

Rule Cards

Multiples of 2	**Multiples of 3**	**Multiples of 4**
Multiples of 5	**Multiples of 6**	**Even numbers**
Odd numbers	**Prime numbers**	**Composite numbers**
Square numbers	**Numbers with 2 as a digit**	**Numbers with exactly 4 factors**

Answer Key for Women's Soccer Results

Answers may vary; see sample responses below.

1.

	X		X	
X	X	X	X	X
0	1	2	3	4

	X				
	X				
X	X	X		X	X
0	1	2	3	4	5

2a.

X						
X						
X	X	X			X	X
0	1	2	3	4	5	6

b.

	X			
	X			X
X	X		X	X
0	1	2	3	4

c.

		X					
X		X					
X	X	X					X
0	1	2	3	4	5	6	7

d.

X				
X			X	X
X			X	X
0	1	2	3	4

3. Students' answers will vary.

4. Two possible data sets include these:

		X					
		X		X			
X		X		X			X
1	2	3	4	5	6	7	8

X					
X					
X		X	X	X	X
2	3	4	5	6	7

References

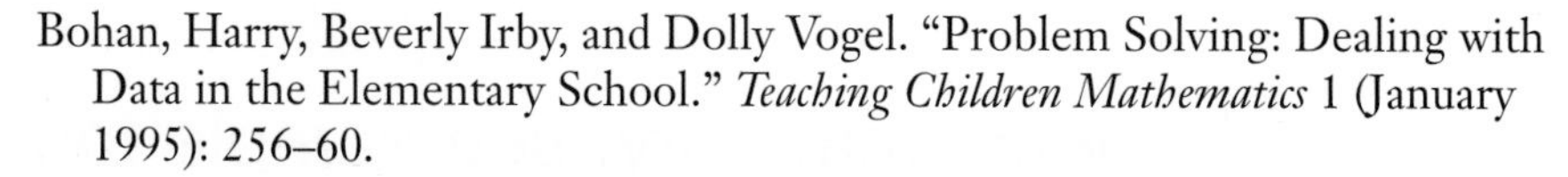
Bohan, Harry, Beverly Irby, and Dolly Vogel. "Problem Solving: Dealing with Data in the Elementary School." *Teaching Children Mathematics* 1 (January 1995): 256–60.

Christensen, Larry B. *Experimental Methodology*. 8th ed. Boston: Allyn & Bacon, 2001.

Cothron, Julia, Ronald Giese, and Richard Rezba. *Science Experiments by the Hundreds*. Dubuque, Iowa: Kendall/Hunt Publishing Co., 1996.

Economopoulos, Karen, Jan Mokros, Rebecca Corwin, and Susan Jo Russell. *From Paces to Feet: Measuring and Data*. Investigations in Number, Data, and Space series. Palo Alto, Calif.: Dale Seymour Publications, 1995.

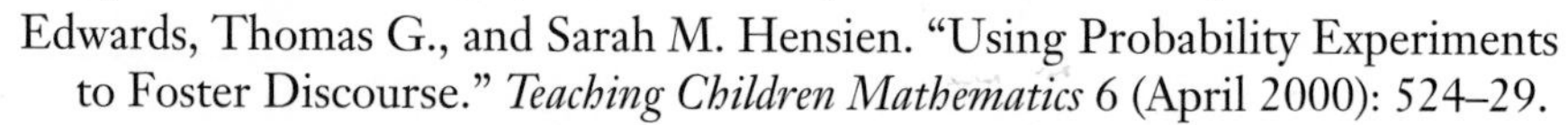
Edwards, Thomas G., and Sarah M. Hensien. "Using Probability Experiments to Foster Discourse." *Teaching Children Mathematics* 6 (April 2000): 524–29.

Feicht, Louis. "Making Charts: Do Your Students Really Understand the Data?" *Mathematics Teaching in the Middle School* 5 (September 1999): 16–18.

Friel, Susan N. "Teaching Statistics: What's Average?" In *The Teaching and Learning of Algorithms in School Mathematics*, 1998 Yearbook of the National Council of Teachers of Mathematics (NCTM), edited by Lorna J. Morrow, pp. 208–17. Reston, Va.: NCTM, 1998.

Friel, Susan N., Janice R. Mokros, and Susan Jo Russell. *Middles, Means, and In-Betweens*. Palo Alto, Calif.: Dale Seymour Publications, 1992.

Garfield, Joan B., and Iddo Gal. "Teaching and Assessing Statistical Reasoning." In *Developing Mathematical Reasoning in Grades K–12*, 1999 Yearbook of the National Council of Teachers of Mathematics (NCTM), edited by Lee V. Stiff, pp. 207–19. Reston, Va.: NCTM, 1999.

Hitch, Chris, and Georganna Armstrong. "Daily Activities for Data Analysis." *Arithmetic Teacher* 41 (January 1994): 242–45.

Jones, Graham A., Carol A. Thornton, Cynthia W. Langrall, and James E. Tarr. "Understanding Students' Probabilistic Reasoning." In *Developing Mathematical Reasoning in Grades K–12*, 1999 Yearbook of the National Council of Teachers of Mathematics (NCTM), edited by Lee V. Stiff, pp. 146–55. Reston, Va.: NCTM, 1999.

Mason, Julia A., and Graham A. Jones. "The Lunch-Wheel Spin." *Arithmetic Teacher* 41 (March 1994): 404–8.

Mokros, Jan, and Susan J. Russell. "Children's Concepts of Average and Representativeness." *Journal for Research in Mathematics Education* 26 (January 1995): 20–39.

Parker, Janet, and Connie C. Widner, eds. "Teaching Mathematics with Technology: Statistics and Graphing." *Arithmetic Teacher* 39 (April 1992): 48–52.

Russell, Susan J., and Jan Mokros. "What Do Children Understand about Average?" *Teaching Children Mathematics* 2 (February 1996): 360–64.

Singer, Margie, Cliff Konold, and Andee Rubin. *Between Never and Always*. Palo Alto, Calif.: Dale Seymour Publications, 1996.

Wiest, Lynda R., and Robert J. Quinn. "Exploring Probability through an Evens-Odds Dice Game." *Mathematics Teaching in the Middle School* 4 (March 1999): 358–62.

Zawojewski, Judith S., and J. Michael Shaughnessy. "Mean and Median: Are They Really So Easy?" *Mathematics Teaching in the Middle School* 5 (March 2000): 436–40.